ORIGO STEPPING STONES 2.0
COMPREHENSIVE MATHEMATICS

AUTHORS
James Burnett
Calvin Irons
Peter Stowasser
Allan Turton

PROGRAM CONSULTANTS
Debi DePaul
Diana Lambdin
Frank Lester, Jr.
Kit Norris

CONTRIBUTING WRITERS
Jaye Kelly
Beth Lewis
Donna Richards

STUDENT BOOK B

ORIGO EDUCATION

CONTENTS

BOOK A

MODULE 1

1.1	Number: Reviewing whole numbers	6
1.2	Number: Reviewing fractions	8
1.3	Number: Reviewing abbreviations for numbers greater than one million	12
1.4	Number: Using exponents greater than 2	14
1.5	Number: Introducing positive and negative numbers	18
1.6	Number: Interpreting the negative symbol	20
1.7	Number: Comparing and ordering positive and negative numbers	24
1.8	Number: Introducing absolute value	26
1.9	Number: Using the 1st and 2nd quadrants of the coordinate plane	30
1.10	Number: Using all quadrants of the coordinate plane	32
1.11	Number: Calculating distance on the coordinate plane	36
1.12	Number: Exploring reflections on the coordinate plane	38

MODULE 2

2.1	Algebra: Reviewing language and conventions	44
2.2	Algebra: Reviewing order of operations	46
2.3	Algebra: Order of operations involving exponents	50
2.4	Algebra: Order of operations involving common fractions and mixed numbers	52
2.5	Number: Reviewing factors and multiples	56
2.6	Number: Finding the least common multiple	58
2.7	Number: Finding the greatest common factor	62
2.8	Algebra: Using the distributive property	64
2.9	Addition/subtraction: Using the standard algorithm (decimal fractions)	68
2.10	Multiplication: Using the standard algorithm (one-digit whole numbers and decimal fractions)	70
2.11	Multiplication: Using the standard algorithm (two-digit whole numbers and decimal fractions)	74
2.12	Multiplication: Using the standard algorithm (decimal fractions)	76

MODULE 3

3.1	Ratio: Introducing ratio	82
3.2	Ratio: Building equivalent ratios pictorially	84
3.3	Ratio: Examining equivalence using tables	88
3.4	Ratio: Exploring ratios on the coordinate plane	90
3.5	Ratio: Calculating and identifying equivalent ratios	94
3.6	Ratio: Interpreting part-part and part-whole situations	96
3.7	Ratio: Solving word problems with part-part and part-whole situations	100
3.8	Division: Reviewing the standard algorithm	102
3.9	Division: Exploring remainders	106
3.10	Division: Terminating and repeating decimal fractions	108
3.11	Division: Adjusting to divide a whole number by a decimal fraction	112
3.12	Division: Adjusting to divide with decimal fractions	114

MODULE 4

4.1	Algebra: Writing expressions to match word problems	120
4.2	Algebra: Writing equations to match word problems	122
4.3	Algebra: Writing equations with two variables	126
4.4	Algebra: Evaluating expressions given the value of the variable	128
4.5	Algebra: Order of operations involving variables	132
4.6	Algebra: Solving equations given a set of possible values	134
4.7	Algebra: Reviewing patterns and rules	138
4.8	Algebra: Interpreting tables	140
4.9	Algebra: Investigating number patterns and rules	144
4.10	Algebra: Exploring different representations of patterns	146
4.11	Algebra: Identifying independent and dependent variables	150
4.12	Algebra: Backtracking to solve equations	152

MODULE 5

5.1	Division: Interpreting division situations	158
5.2	Division: Common fractions (same denominators)	160
5.3	Division: Common fractions (related denominators)	164
5.4	Division: Whole numbers by common fractions	166
5.5	Division: Whole numbers by common fractions (with remainders)	170
5.6	Division: Common fractions by common fractions (unrelated denominators)	172
5.7	Division: Consolidating strategies	176
5.8	Ratio: Comparing ratios in tables	178
5.9	Ratio: Comparing ratios in tables and graphs	182
5.10	Ratio: Using a given ratio when the total is known	184
5.11	Ratio: Using a given ratio when total is unknown	188
5.12	Ratio: Working with measurement	190

MODULE 6

6.1	Area: Exploring parallelograms	196
6.2	Area: Using a formula for parallelograms	198
6.3	Area: Exploring right triangles	202
6.4	Area: Using a formula for triangles with height inside	204
6.5	Area: Using a formula for triangles with height inside or outside	208
6.6	Area: Calculating the area of any quadrilateral	210
6.7	Area: Calculating the area of any polygon	214
6.8	Ratio: Developing the concept of rate	216
6.9	Ratio: Identifying rates	220
6.10	Ratio: Rates with whole numbers and fractions	222
6.11	Ratio: Working with rates in two directions	226
6.12	Ratio: Comparing rates	228

STUDENT GLOSSARY AND TEACHER INDEX   234

CONTENTS

BOOK B

MODULE 7

7.1	Algebra: Simplifying expressions	244
7.2	Algebra: Simplifying expressions using the commutative and associative properties	246
7.3	Algebra: Simplifying expressions using the distributive property	250
7.4	Algebra: Simplifying expressions with more than one variable	252
7.5	Algebra: Introducing balance to solve addition equations	256
7.6	Algebra: Solving addition equations	258
7.7	Algebra: Solving subtraction equations	262
7.8	Algebra: Solving multiplication equations	264
7.9	Algebra: Solving division equations	268
7.10	Algebra: Solving word problems (addition and multiplication)	270
7.11	Algebra: Solving word problems (subtraction and division)	274
7.12	Algebra: Solving word problems (all operations)	276

MODULE 8

8.1	Ratio: Linking part-whole ratios to fractions	282
8.2	Ratio: Relating fraction representations of ratio	284
8.3	Ratio: Introducing percentage (area model)	288
8.4	Ratio: Consolidating percentage (number line model)	290
8.5	Ratio: Simple percentages of quantities	294
8.6	Ratio: Percentages of collections	296
8.7	Ratio: Percentages of numbers less than ten	300
8.8	Ratio: Using unit percentages	302
8.9	Ratio: Finding the whole given a part and the percentage	306
8.10	Division: Introducing the invert-and-multiply method (common fractions)	308
8.11	Division: Consolidating the invert-and-multiply method (common fractions)	312
8.12	Division: Consolidating strategies (common fractions)	314

MODULE 9

9.1	Algebra: Reviewing inequalities	320
9.2	Algebra: Showing inequalities on a number line	322
9.3	Algebra: Identifying the range of possible values for an inequality	326
9.4	Algebra: Working with inequalities	328
9.5	Statistics: Introducing statistics	332
9.6	Statistics: Identifying the mode	334
9.7	Statistics: Calculating the median	338
9.8	Statistics: Calculating the mean	340
9.9	Area: Using nets to calculate surface area of prisms	344
9.10	Area: Using nets to calculate surface area of pyramids	346
9.11	Area: Calculating surface area of prisms and pyramids	350
9.12	Area: Solving word problems	352

MODULE 10

10.1	Statistics: Measuring variability using mean and absolute deviation	358
10.2	Statistics: Measuring variability using quartiles and interquartile range	360
10.3	Statistics: Introducing box plots	364
10.4	Statistics: Consolidating box plots	366
10.5	Statistics: Introducing histograms	370
10.6	Statistics: Analyzing and creating histograms	372
10.7	Statistics: Working with histograms	376
10.8	Volume: Reviewing volume	378
10.9	Volume: Rectangular-based prisms with one fractional side length	382
10.10	Volume: Rectangular-based prisms with two fractional side lengths	384
10.11	Volume: Rectangular-based prisms with three fractional side lengths	388
10.12	Volume: Solving word problems	390

MODULE 11

11.1	Ratio: Introducing ratios with three parts	396
11.2	Ratio: Using ratios with three parts	398
11.3	Ratio: Comparing ratios with three parts	402
11.4	Ratio: Solving word problems with three-part ratios	404
11.5	Ratio: Resizing 2D shapes to a given percent	408
11.6	Ratio: Examining similar rectangles	410
11.7	Ratio: Examining similar triangles	414
11.8	Ratio: Examining percentage changes of area	416
11.9	Ratio: Working with percentage changes of area	420
11.10	3D objects: Analyzing pyramid nets	422
11.11	3D objects: Analyzing prism nets	426
11.12	3D objects: Creating prism nets	428

MODULE 12

12.1	Ratio: Introducing percentages greater than 100%	434
12.2	Ratio: Consolidating percentages greater than 100%	436
12.3	Ratio: Using complementary percentages	440
12.4	Ratio: Using percentages greater than 100%	442
12.5	Algebra: Simplifying expressions involving percentages	446
12.6	Algebra: Solving word problems involving percentages	448
12.7	Algebra: Solving equations with percentages and variables	452
12.8	Algebra: Solving word problems with percentages and variables	454
12.9	Algebra: Generating and graphing variables	458
12.10	Algebra: Generating and graphing variables (non-equivalent ratios)	460
12.11	Algebra: Generating and graphing variables (approximate ratios)	464
12.12	Algebra: Generating and graphing variables (non-linear)	466

STUDENT GLOSSARY AND TEACHER INDEX 474

ORIGO Stepping Stones • Grade 6

7.1 Algebra: Simplifying expressions

Step In

What is the area of this square tile? How do you know?

What name can we give to the tile?

What is the area of this tile? How do you know?
What name can we give to this tile?

What do you know about the area of this tile?
What name can we give to this tile?

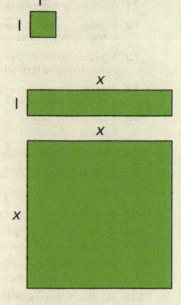

> The small square has an area of one, so I'll call it one. I don't know the length of the long side of the rectangle, so I'll name the area $1x$ or just x. The area of the large square then becomes x^2.

What expression can we write to name this collection of tiles?

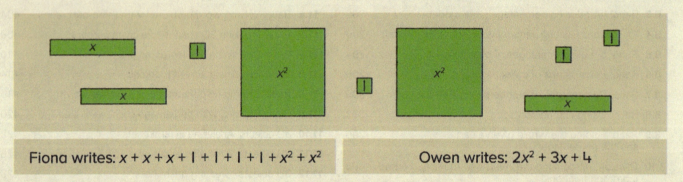

Fiona writes: $x + x + x + 1 + 1 + 1 + 1 + x^2 + x^2$

Owen writes: $2x^2 + 3x + 4$

How does each expression match the collection of tiles?

Are the two expressions equivalent? How do you know?

Step Up

1. Write an expression to name each collection of tiles.

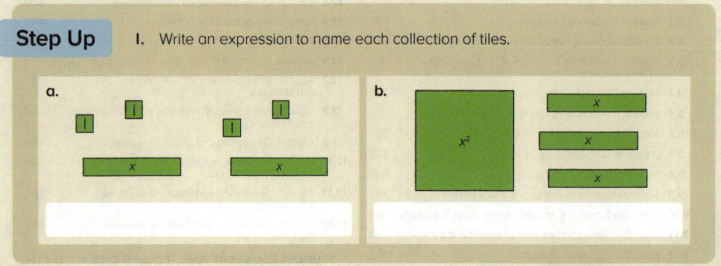

a.

b.

2. Write an expression to name each collection of tiles. Try to write each expression with as few terms as possible.

a.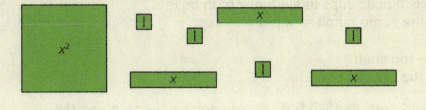

b.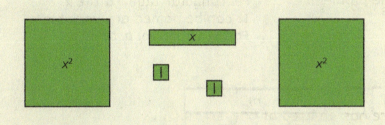

3. Draw a picture of tiles to match each expression.

a. 5x + 3

b. x^2 + 4x + 2

Step Ahead Expand each expression below to show it a different way.

a. 3x + 2
2x + x + 1 + 1

b. 4 + 2x

c. $2x^2$ + 3x + 1

d. 7 + $3x^2$ + x

e. 3x + x^2 + 4x

7.2 Algebra: Simplifying expressions using the commutative and associative properties

Step In

A party store sells plastic cups individually or in boxes. Each box holds the same number of plastic cups.

There are 5 cups and 1 box on the top shelf.
There are 4 cups and 2 boxes on the bottom shelf.

Let m represent the number of plastic cups in each box. What picture could you draw to model the problem?

> You don't have to name the rectangular algebra tile x. It can be named any variable. For example, m, p, or h.

Charlotte draws this picture of algebra tiles.

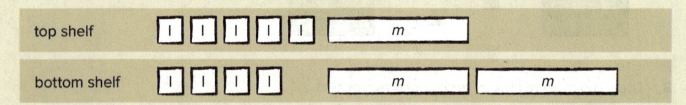

How does her picture model the problem? What expression could she write to represent the total number of plastic cups?
The expressions below all match the picture. Which is easiest to read?

A $5 + m + 4 + 2m$

B $5 + 4 + m + 2m$

C $1 + 1 + 1 + 1 + 1 + m + 1 + 1 + 1 + 1 + m + m$

To **simplify** the expression further, you combine or collect **like terms**. This means all single numbers are combined, all the same variables are combined, and all the same squared variables are combined. For the example above, the result would be $9 + 3m$, or $3m + 9$.

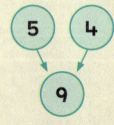
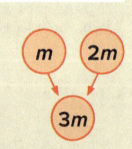

Step Up

1. a. Write an expression for the tiles on the left and an expression for the tiles on the right.

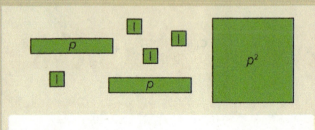

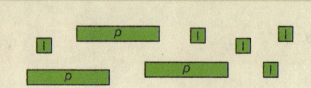

b. Combine the two expressions into a single expression to represent the total amount represented by the tiles.

2. Write an expression to represent the total amount.

a.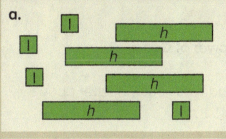

b.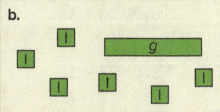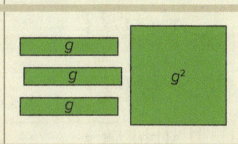

3. Draw a picture of tiles to match the expression. Then simplify into an equivalent expression.

a. $c + 8 + c + c =$

b. $4k + 3 + 2k =$

c. $d + d + 3 + d + d^2 =$

d. $b + b + 5 + 3b =$

Step Ahead

Write an expression to show how you would solve this problem. Let t represent the number of *tulips* in each bunch. Then draw a picture to model the problem.

Tulips are bought individually or in a bunch. Each bunch has the same number of tulips. There are 5 bunches of tulips sold on Friday. On Sunday, there are 3 bunches of tulips and 6 individual tulips sold. How many tulips are sold in total for both days?

7.2 Maintaining concepts and skills

Computation Practice

What is the official state wildflower of Georgia?

★ Complete each equation. Find each product in the grid below and cross out the letter above. Then write the remaining letters at the bottom of the page.

12 × $2.10 = $ ___	$1.20 × 16 = $ ___	8 × $1.30 = $ ___
$2.50 × 18 = $ ___	14 × $2.70 = $ ___	$3.50 × 22 = $ ___
6 × $1.40 = $ ___	$2.20 × 8 = $ ___	12 × $1.70 = $ ___
$1.30 × 18 = $ ___	16 × $2.40 = $ ___	$2.10 × 14 = $ ___
12 × $3.50 = $ ___	$1.50 × 18 = $ ___	22 × $1.30 = $ ___
$2.80 × 16 = $ ___	8 × $2.70 = $ ___	$1.70 × 6 = $ ___
14 × $2.20 = $ ___	$2.30 × 12 = $ ___	6 × $2.60 = $ ___
$1.50 × 24 = $ ___	36 × $2.50 = $ ___	$1.40 × 8 = $ ___

C	N	A	A	L	Z
$37.80	$20.40	$27.60	$92.00	$25.20	$37.60
T	H	E	S	B	O
$42.00	$44.80	$77.00	$38.40	$23.40	$15.60
A	A	S	T	E	N
$28.60	$25.40	$29.40	$30.80	$36.00	$45.00
L	S	C	L	E	V
$27.80	$11.20	$10.40	$21.60	$20.60	$8.40
E	F	S	A	A	S
$10.20	$27.00	$19.20	$90.00	$29.60	$17.60

Write the letters in order from the ✱ to the bottom-right corner.

Ongoing Practice

1. Read each story and write a matching rate.

a. Plastic cups cost $4 for 20. What rate describes the price for 1 cup?

b. Cathy paid $56 for 8 tickets. What rate describes the price of 1 ticket?

c. Allan uses 8 eggs to bake 4 cakes. What rate describes the number of eggs in 1 cake?

d. Twenty pencils cost $5. What rate describes the price of 1 pencil?

2. Write an expression to name each collection of tiles. Try to write each expression in its simplest form.

a.

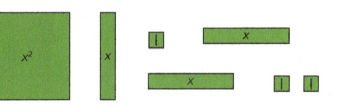

b.

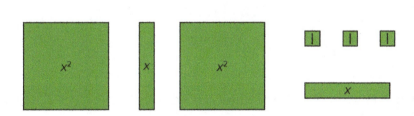

Preparing for Module 8

Identify whether each question involves a part-part or part-whole situation.

a. Hugo's class has 28 students. There are 15 boys in his class. What is the ratio of boys to girls?
- ○ part-part
- ○ part-whole

b. Jennifer bought 5 bananas and 8 apples. What is the ratio of bananas to apples?
- ○ part-part
- ○ part-whole

c. Awan has a collection of 56 basketball cards. Of these, 18 are special edition cards. What is the ratio of special edition cards to the total number of cards?
- ○ part-part
- ○ part-whole

7.3 Algebra: Simplifying expressions using the distributive property

Step In Carmen makes three bracelets. There are five orange beads on each bracelet. The rest of the beads are green. Each bracelet has an equal number of beads.

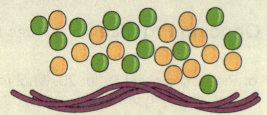

Let g represent the number of green beads.
What picture could you draw to model the story?

Cary draws this picture of algebra tiles.

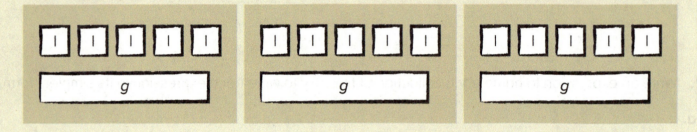

How does the picture represent the story?

What expression would you write to match the picture and word story?

What do you notice about each of these expressions?

| $5 + 5 + 5 + g + g + g$ | $(3 \cdot 5) + (3 \cdot g)$ | $3(5 + g)$ | $15 + 3g$ |

What is the same about each expression? How do you know they are equivalent?

Jacinta has a room she needs to carpet. Some furniture is currently in the way so she cannot measure the dimensions completely. She knows the width is 3 yards and the length that she can reach is 5 yards. She draws this picture to match what she knows so far about the area of the room.

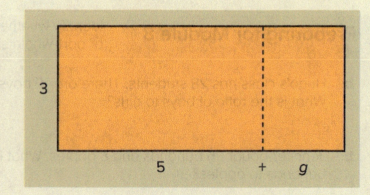

Write an expression to represent the story.

How does the expression compare to the expressions written for the bead story above?

Step Up 1. Shade the expressions that match the picture.

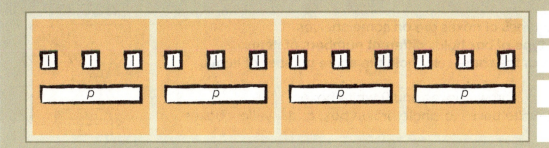

4 (p + 3)

(4 • p) + (4 • 3)

12 + 4p

4p + 3

2. Write an expression to match each problem.

a. There are 5 equal rows of fruit trees. In each row, there are 15 apple trees and some plum trees, with the same number of plum trees in each row. Let p represent the number of plum trees in each row. How many fruit trees are there in total?

b. Each concert ticket costs $49 plus a $5 service fee. Ethan buys a ticket for himself and a group of friends. Let n represent the number of tickets that he bought. What is the total cost of the tickets?

c. A garden path is $1\frac{1}{2}$ yards wide. The first 10 yards of the path are concrete. The rest of the path is gravel. Let p represent the length in yards of the path that is gravel. What is the total area of the path?

3. Simplify each expression.

a. $3(m + 4) =$

b. $7(5 + d) =$

c. $h(8 + 2) =$

d. $4(1 + 2b) =$

e. $15(a + 7) =$

f. $j(1 + 6) =$

Step Ahead   Draw a picture of algebra tiles to match each expression.

a. $3(h + 4)$

b. $3(2m + 4)$

7.4 Algebra: Simplifying expressions with more than one variable

Step In

Three types of boxes are on some shelves.
Each type of box holds different numbers of items.
Boxes of the same color hold the same number of items.

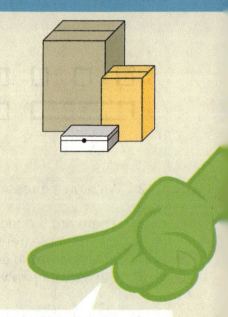

On the top shelf there are 3 brown boxes and 9 white boxes.
The bottom shelf has 2 white boxes, a single brown box, and 4 yellow boxes.

Let b represent the number of items in each brown box,
w the number in each white box, and y the number in each yellow box.
Write expressions to represent the total number of items on each shelf.

Combine like terms to write an expression that represents the total number of each type of box on both shelves.

I need to remember to keep terms separate. I can't take $9w$ and $3b$ and make $9 + 3 + w + b$, or $12 + wb$. But I can take $9w$ and $2w$ and make $11w$.

Emily prints 18 pages in color and 27 pages in black and white.
It costs more to print a page in color.

To show the total cost, Juan writes $18c + 27b$ where c represents the cost of printing a page in color and b the cost of a black and white page.

The greatest common factor that the two terms share is 9, so I could write $9(2c + 3b)$.

Why are these expressions equivalent?

What other equivalent expression could you write?
How do you know?

Step Up

1. Write each expression so that like terms are grouped together.

a. $3a + 8c + 1a + 5 =$

b. $9 + x + 5x + 4 + y =$

c. $14 + g + g + 6 + g^2 =$

d. $6k + 7 + d^2 + 20k + 0 =$

e. $m + 9m^2 + m + 2m =$

f. $12d + 7f + 2g + 3f + d^2 =$

2. Write the greatest common factor (GCF) that each term shares.

a. 15d + 10e
The GCF is ☐

b. 9k + 36h
The GCF is ☐

c. 16x + 12j
The GCF is ☐

d. 30b + 45g
The GCF is ☐

e. 8c + 10d + 4c
The GCF is ☐

f. 12v + 6h + 30h
The GCF is ☐

g. 32e + 20 + 4e
The GCF is ☐

h. 12 + 30h + 24
The GCF is ☐

3. Find the greatest common factor. Then rewrite each expression.

a. 27h + 15g =
3 (9 h + 5 g)

b. 16f + 40g =
☐ (☐ f + ☐ g)

c. 20d + 35 =
☐ (☐ d + ☐)

d. 36 + 18b =
☐ (☐ + ☐)

e. 45j + 60k =
☐ (☐ + ☐)

f. 32w + 24z =
☐ (☐ + ☐)

4. Write an expression to match each problem.

a. Nam buys nine bags of red apples and five bags of green apples. Each bag has the same number of apples. How many apples are there in total? Let r represent the number of red apples in each bag and g represent the number of green apples in each bag.

b. Gasoline costs $2.34 per gallon. Michelle fills up her car then fills up a large fuel can to take with her. What is the total amount she pays? Let c represent the number of gallons that fill the car and let f represent the number of gallons that fill the can.

Step Ahead Cross out the expression that is not equivalent. Show your thinking.

a. (6 • b) + (24 • c)

6b + 24c
6 (b + 4c)
3 (2b + 8c)
6b • 24c

b. 60p + 12m

12 (5p + m)
(3 • 20p) + (3 • 6m)
6 (10p + 2m)
4 • 15p + 4 • 3m

7.4 Maintaining concepts and skills

Think and Solve

What is the mass of each box in whole kilograms?

Clues
- A is twice the mass of B.
- B is twice the mass of C.
- C is twice the mass of D.
- All 4 together weigh exactly 90 kg.

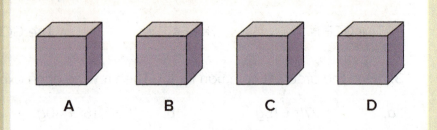

A ☐ kg   B ☐ kg   C ☐ kg   D ☐ kg

Words at Work

Research and write about how algebra is used in everyday life.

Ongoing Practice

1. Complete the two rates to match each story.

a. Kayla buys 2 lb of apples for $8.

The rate is $ ☐ for 1 lb.

The rate is ☐ lb for $1.

b. Corey uses 9 fl oz of fruit concentrate for every 3 pt of water.

The rate is ☐ fl oz for 1 pt.

The rate is ☐ pt for every 1 fl oz.

c. A dressmaker spends $60 on 12 yards of fabric.

The rate is $ ☐ for 1 yd.

The rate is ☐ yd for $1.

d. A banana bread recipe has 4 bananas for every 2 cups of flour.

The rate is ☐ bananas for 1 cup.

The rate is ☐ cup for every 1 banana.

2. Draw a picture to match the expression. Then write an equivalent expression to simplify it.

a. $d + d + d + 4 =$

b. $3v + 8 + 2v =$

c. $3e + 5 + 2e + e + e =$

Preparing for Module 8

Write the product of each of these first as an improper fraction, and then as a mixed or whole number.

a. $6 \times \frac{2}{5} =$ ☐ = ☐

b. $5 \times \frac{3}{4} =$ ☐ = ☐

c. $8 \times \frac{7}{8} =$ ☐ = ☐

d. $5 \times \frac{2}{6} =$ ☐ = ☐

e. $9 \times \frac{2}{3} =$ ☐ = ☐

f. $7 \times \frac{6}{10} =$ ☐ = ☐

7.5 Algebra: Introducing balance to solve addition equations

Step In

Algebra tiles can be used to model and solve equations.

What equation could you write to match this pan balance picture?

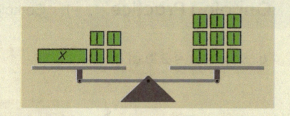

The two sides balance each other. The small squares have a value of one, so I could write $x + 4 = 9$.

How could you determine the value of x?

Blake crosses out the same number of small tiles on each side of the pan balance.
How did he decide what number of tiles to cross out?

What is the value of x? How could you check?

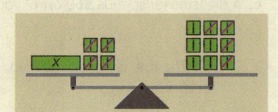

If I substitute x for the number of blocks that I think it is worth I should have the same amount on each side.

Step Up

1. Write an equation to match each pan balance.

a.
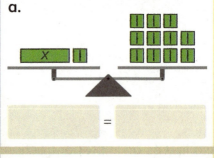
=

b.
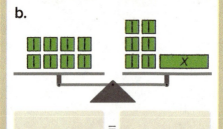
=

c.
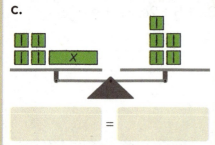
=

d.
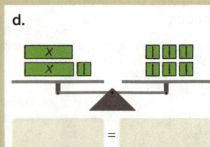
=

e.
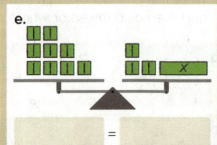
=

f.
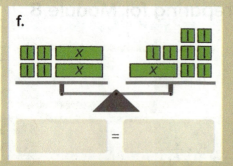
=

2. Write an equation to match the picture. Then calculate the value of the variable. Cross out algebra tiles to show your thinking.

a. If $z + 4$ =

Then z =

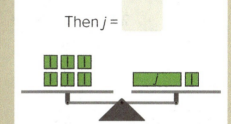

b. If =

Then p =

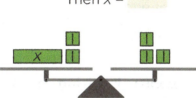

c. If =

Then m =

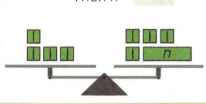

d. If =

Then j =

e. If =

Then x =

f. If =

Then n =

3. Draw algebra tiles on the pan balance to match the equation. Then cross them out to show the solution.

a. If $x + 2 = 7$

Then x =

b. If $14 = y + 6$

Then y =

Step Ahead Calculate the value of each variable. Show your thinking.

a. If $m + 4 = 7.5$

Then m =

b. $11 = w + 5.2$

Then w =

7.6 Algebra: Solving addition equations

Step In

What equation could you write to match this pan balance?

What steps would you follow to calculate the value for *g*?

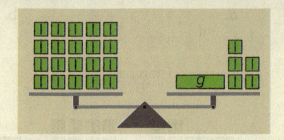

> I would cross out five ones on each side of the pan balance.

Fatima writes these steps to show her thinking.
What steps does she follow?

How does the equation remain balanced at each step?

How can you check the value of *g* is correct?

$$20 = g + 5$$
$$20 - 5 = g + 5 - 5$$
$$15 = g$$

Kevin uses Fatima's method to solve a problem involving common fractions.

What advantages does this method have over drawing a pan balance?

For what other types of numbers would this method be more useful than drawing a pan balance?

$$\frac{17}{6} = h + \frac{3}{6}$$
$$\frac{17}{6} - \frac{3}{6} = h + \frac{3}{6} - \frac{3}{6}$$
$$\frac{14}{6} = h$$

What steps would you take to solve equations like this?
$41 = 5 + m + 6$

> I think I'd start by collecting like terms instead of doing extra subtraction.

Step Up

1. Write an equation to match the picture. Cross out blocks to determine the value of the variable. Then use Fatima's method to show the calculations.

a.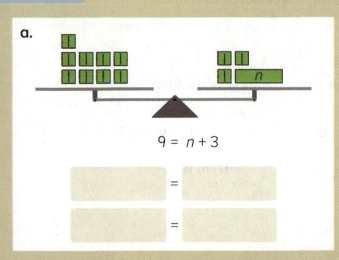

$9 = n + 3$

b.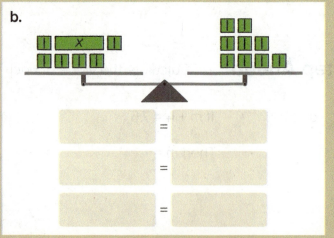

2. Calculate the value of each variable. Show your thinking and be sure to show each step.

a. $p + 3 = 21$

$p + 3 - 3 = 21 - 3$

b. $41 = m + 1$

c. $65 = a + 5$

d. $w + 1.5 = 4.2$

e. $z + \frac{12}{20} = 2\frac{6}{20}$

f. $48 + 3 = p + 30$

g. $z + 1.20 + 0.8 = 7.8$

h. $58 + 34 = 26 + p + 30$

3. Manuel, Kylie, and Helen play a game. Manuel scores 43 points and Helen scores 36. The total score for all three friends is 98 points. What is Kylie's score? Let b represent Kylie's score. Show your thinking.

Step Ahead

Mika tried solving the equation shown on the right. Explain what he did wrong, then show a correct solution on page 280.

$n + \frac{5}{6} + \frac{5}{2} = \frac{28}{6}$

$n + \frac{10}{6} = \frac{28}{6}$

$n + \frac{10}{6} - \frac{10}{6} = \frac{28}{6} - \frac{10}{6}$

$n = \frac{18}{6}$

7.6 Maintaining concepts and skills

Computation Practice

★ Complete the equations. Then write each letter above its matching answer at the bottom of the page to discover a science fact. Some letters are used more than once.

$\frac{1}{4}$ of 240 = ___ **v**	$\frac{1}{5}$ of 160 = ___ **i**	$\frac{1}{10}$ of 720 = ___ **f**
$\frac{1}{4}$ of 440 = ___ **a**	$\frac{1}{5}$ of 270 = ___ **b**	$\frac{1}{10}$ of 260 = ___ **u**
$\frac{1}{4}$ of 320 = ___ **t**	$\frac{1}{5}$ of 340 = ___ **h**	$\frac{1}{10}$ of 560 = ___ **s**
$\frac{1}{4}$ of 180 = ___ **y**	$\frac{1}{5}$ of 420 = ___ **d**	$\frac{1}{10}$ of 640 = ___ **q**
$\frac{1}{4}$ of 340 = ___ **o**	$\frac{1}{5}$ of 530 = ___ **c**	$\frac{1}{10}$ of 190 = ___ **r**
$\frac{1}{4}$ of 360 = ___ **w**	$\frac{1}{4}$ of 220 = ___ **e**	

☐ ☐ ☐ ☐
85 60 55 19

☐ ☐ ☐ ☐ ☐ - ☐ ☐ ☐ ☐ ☐ ☐ ☐
80 68 19 55 55   64 26 110 19 80 55 19 56

☐ ☐   ☐ ☐ ☐   ☐ ☐ ☐ ☐ ☐ ' ☐
85 72   80 68 55   55 110 19 80 68   56

☐ ☐ ☐ ☐ ☐ ☐ ☐ ☐ ☐
56 26 19 72 110 106 55   32 56

☐ ☐ ☐ ☐ ☐ ☐ ☐ ☐ ☐
106 85 60 55 19 55 84   54 45

☐ ☐ ☐ ☐ ☐
90 110 80 55 19

Ongoing Practice

1. Identical products are being sold. Shade the ○ beside the better offer. Show your thinking.

a. ○ 6 muffins for $9   ○ 4 muffins $5

b. ○ 8 tickets for $320   ○ 6 tickets for $270

c. ○ 4 cards for $8   ○ 8 cards for $12

d. ○ 5 kg for $15   ○ 3 kg for $12

2. Write each expression so that like terms are grouped together.

a. $4d + 2 + 2d + 5e =$

b. $20x + 20x^2 + x =$

c. $15f + f + f + f + 5 + 2g =$

d. $4p + p + 4 + 21m^2 =$

e. $b + b + b + 3c + 2c + 5 =$

f. $v + 5v^2 + v + 2v =$

Preparing for Module 8

Each large rectangle below is one whole. Complete the sentence and color the diagram to show each share.

a. $\frac{2}{3}$ is equivalent to ☐ divided by ☐

b. $\frac{3}{7}$ is equivalent to ☐ divided by ☐

7.7 Algebra: Solving subtraction equations

Step In

Deon's lunch costs $9. He now has $15 left in his wallet. How much money did he have in his wallet before lunch?

What equation would you write to represent this problem? Let m represent the unknown value.

What operation would you use to solve the problem?

What pan balance picture could you draw to model it?

It would be hard to draw a picture. I would need to show what the pan balance looks like before the 9 ones are taken away from m.

Julia follows these steps to calculate the value of m.

Why does she add nine to each side of the equation?
How can she check that the value for m is correct?

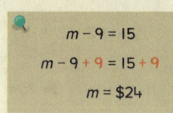

$$m - 9 = 15$$
$$m - 9 + 9 = 15 + 9$$
$$m = \$24$$

How could you apply the same thinking to calculate $18 = h - 3$?

Step Up

1. Write an equation to represent the picture. Let n represent the unknown value. Then calculate the value of n.

Garbo's Grill

Cayo Taco	$11.50
Kogi Dog	$8.00
Subtotal	$19.50
Sales Tax	$
Total	$20.96

262

2. Calculate the value of each variable. Show your thinking.

a. $m - 16 = 46$

$m - 16 + 16 = 46 + 16$

b. $65 = p - 21$

c. $32 = s - 28$

d. $k - 2.7 = 1.2$

e. $g - 39 = 53$

f. $3\frac{6}{8} = q - 4\frac{4}{8}$

3. Write an equation to represent each problem. Then calculate the value of the variable. Show your thinking.

a. Fifteen minutes have passed since a baker put a cake into the oven. The kitchen timer says that the cake must bake for another twenty-five minutes. What was the original baking time?
Let b represent the original baking time.

b. Hannah buys some pizzas for a party. Two and half pizzas are eaten. There are one and a half pizzas left. How many pizzas did Hannah buy for the party? Let p represent the number of pizzas she bought.

Step Ahead

Find the value of each variable. Show your thinking.

a. $g - 5 + 8 = 15$

b. $60 = n + 42 - 9$

7.8 Algebra: Solving multiplication equations

Step In Megan buys three plants for $18. Each plant costs the same amount. What is the cost of one plant?

What equation could you write to solve the problem? Let f represent the cost of each plant.

I know that $3f = 18$, but I need to figure out the value of f.

What steps would you use to solve the problem? What operation would you use?

What pan balance could you draw to model it?

Hunter draws this picture.
How does it match the problem?

What is the value of f? How do you know?

The pans stay balanced if I remove one-third of each amount, then remove the same amount again. Why does this happen?

Andrea follows these steps instead of drawing a picture.
Why does she divide by $3f$ by 3?
Why doesn't she divide by $3f$ instead?

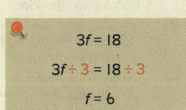

$3f = 18$
$3f \div 3 = 18 \div 3$
$f = 6$

Carter is evaluating the equation $6n \cdot 4 = 48$.
He begins writing steps to show his thinking.
What step could he make next?
What other step could he take?

$6n \cdot 4 = 48$
$6 \cdot n \cdot 4 = 48$

How would you solve the equation $5c + 3c + c = 63$?

I could try dividing both sides by 5 to keep the equation balanced. But that won't give me the easiest amount to work with.

I think it'd make sense to combine like terms before I do anything else. $5c$ and $3c$ is $8c$, then one more c is $9c$. Now I'm ready to solve the equation.

Step Up

1. Calculate the value of each variable. Show your thinking.

a. $2d = 16$

$2d \div 2 = 16 \div 2$

b. $27 = 9y$

c. $3h = 4.5$

d. $32 + 24 = 7r$

e. $55 = 4g + 7g$

f. $z + 8z = 4 + 32$

g. $5\frac{2}{10} = 4g$

h. $2z + z + 5z = 4$

2. Write a word problem to match the equation. $24r = 4$

Step Ahead

Complete the number trail.

12 → +5 → 17 → +3n → ☐ → −5 → ☐ → +5n → ☐ → ☐ → 4 + 8n → ☐ → 4 + 7n

7.8 Maintaining concepts and skills

Think and Solve

In the school band, the ratio of boys to girls is 4 to 5. There are three more girls than boys in the band.

a. How many students are in the school band?

b. Write how you figured it out.

Words at Work

Write a word problem to match the equation. Then calculate the value of the variable.

$52 + 43 + x = 142$

Ongoing Practice

1. Complete each table to show the dimensions of four different prisms that have the same volume.

a.
Volume is 36 m³		
Length	Width	Height

b.
Volume is 90 in³		
Length	Width	Height

2. Write an equation to match each pan balance.

a.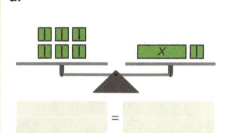

=

b.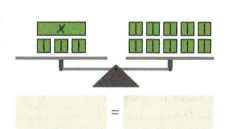

=

c.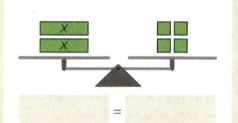

=

d.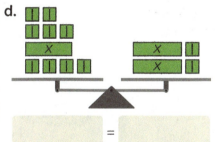

=

e.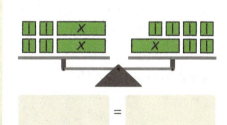

=

f.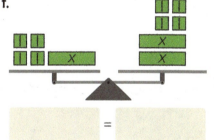

=

Preparing for Module 8

Write each of these as a multiplication equation. Then complete the equation.

a. $\frac{1}{3}$ of 27 =

b. $\frac{1}{5}$ of 18 =

c. $\frac{1}{6}$ of 24 =

7.9 Algebra: Solving division equations

Step In

If you divide a number by 5, you get a quotient of 7.

What equation would you write to model the problem? Let y represent the mystery number.

A quantity divided by 5 will leave 7 in each group. I'll write $y \div 5 = 7$.

Karen draws a picture.

How does the picture help to calculate the value of y?

Liam calculates the value of y like this.

$y \div 5 = 7$

$y \div 5 \cdot 5 = 7 \cdot 5$

$y = 35$

What steps does he follow?
Why does he multiply each side of the equation by 5?
How can you prove the value for y is correct?

You can substitute 35 in for y in the original equation and see if each side of the equal symbol is the same amount.

How could you use Liam's strategy to evaluate $p \div 3 = 15$?

Step Up

1. Calculate the value of each mystery number. Use Liam's strategy to show your thinking.

    a. If you divide a number by 9, the quotient is 6. Let m represent the mystery number.

    b. If you divide a number by 15, the quotient is 3. Let h represent the mystery number.

2. Calculate the value of each mystery number. Show your thinking.

a. If you divide a number by 5, the quotient is 12. Let z represent the mystery number.

b. A number divided by $\frac{1}{8}$ is 11. Let m represent the mystery number.

3. Calculate the value of each variable. Show your thinking.

a. $b \div 5 = 30$

b. $15 = p \div 4$

c. $y \div 10 = 2.65$

d. $r \div \frac{1}{5} = 35$

e. $a \div 8 = 22 - 7$

f. $45 \div 9 = e \div 2.5$

Step Ahead Calculate the value of each variable. Show your thinking.

a. $2w \div 6 = 12$

b. $4k \div 8 = 16$

7.10 Algebra: Solving word problems (addition and multiplication)

Step In

How would you solve this problem?

Menu	
Sub	Cost
Turkey	$8.95
Ham	$7.50
Tuna	$6.95
Chicken	$7.95

Terri has some money in her purse. She borrows $3 from her sister and another $2 from her best friend to buy a turkey sub. No change is given. How much money did she have before?

I can calculate the solution in my head. I'll let p represent the amount of money that she had before. That's $p + 3 + 2 = 8.95$.

Even though you can calculate the answer in your head, it can be helpful to think about the problem like this.

$p + 3 + 2 = 8.95$
$p + 5 = 8.95$
$p + 5 - 5 = 8.95 - 5$
$p = 3.95$

What steps are followed?

What like terms are combined?

Why is 5 subtracted from each side of the equation?

How does each step stay balanced?

Ramon orders three of the same type of sub.

The total is $22.50. What type of sub did he order? Let a represent the type of sub that he ordered.

$3a = 22.50$
☐ ÷ 3 = ☐ ÷ 3
$a = $ ☐

Write the missing numbers. Then describe the steps that are followed.

Step Up

1. Use the prices at the top of the page to solve this problem. Show your thinking.

Samuru buys a ham sub. He has $3.80 in cash. He uses his debit card to pay the remainder. Calculate the amount that Samuru paid on his debit card.

Let d represent the unknown amount.

$d + $ ☐ $= $ ☐
$d + $ ☐ $- $ ☐ $= $ ☐ $- $ ☐
$d = $ ☐

2. Write an equation to match each problem. Then solve the equation and write the answer below. Remember to include the appropriate unit.

a. A machine assembles 60 bicycles in 4 hours. A person assembles 60 bicycles in 9 hours. How many bicycles does the machine assemble each hour? Let b represent the number of bicycles.

b. Three friends enter a 15-mile fun run. Dwane runs the first $4\frac{1}{2}$ miles. Natalie runs the next $3\frac{1}{4}$ miles. Reece runs the rest of the distance. How far did Reece run? Let r represent the unknown distance.

c. Concert tickets cost $15 plus a $2 service fee. Gloria buys one ticket but only has $9.50. Her mother gives her the extra money. How much money did her mother contribute? Let m represent the amount her mother contributes.

d. There are 6 DVDs in a boxed set. Luke pays $60 and his mother pays $12 to buy the set. They receive no change. What is the cost of each DVD? Let d represent the cost of each DVD.

Step Ahead

Write an equation to represent the problem. Then solve the problem and write the answer below.

Logan rents a car for 4 days. He is charged an additional $37 because he fails to fill the gas tank before returning the car. He is charged a total of $372. What was the original cost of renting the car each day? Let h represent the cost of renting the car each day.

7.10 Maintaining concepts and skills

Computation Practice

★ Complete the equations. Then write each letter above its matching difference at the bottom of the page to discover a fact about the human body. Some letters are used more than once.

15 − 5.4 = __ h	30 − 5.6 = __ t	12 − 3.4 = __ o
29 − 3.1 = __ i	16 − 5.7 = __ n	32 − 3.5 = __ e
35 − 6.7 = __ b	20 − 3.8 = __ e	20 − 7.2 = __ c
28 − 4.5 = __ g	14 − 3.5 = __ r	44 − 3.8 = __ o
54 − 8.1 = __ s	15 − 3.9 = __ a	19 − 1.3 = __ n
48 − 7.5 = __ h	24 − 2.3 = __ g	21 − 1.8 = __ t
55 − 2.6 = __ e	25 − 4.6 = __ e	35 − 1.8 = __ n
22 − 8.3 = __ s	14 − 5.5 = __ i	22 − 2.4 = __ h

☐ ☐ ☐
24.4 19.6 52.4

☐ ☐ ☐ ☐ ☐
19.2 40.5 25.9 23.5 40.5

☐ ☐ ☐ ☐
28.3 40.2 17.7 20.4

☐ ☐
8.5 45.9

☐ ☐ ☐ ☐ ☐ ☐ ☐
13.7 24.4 10.5 40.2 33.2 21.7 52.4 10.5

☐ ☐ ☐ ☐
19.2 9.6 11.1 10.3

☐ ☐ ☐ ☐ ☐ ☐ ☐
12.8 8.6 10.3 12.8 10.5 28.5 19.2 16.2

Ongoing Practice

1. Calculate the volume of each prism. Show your thinking. Remember to include the appropriate unit.

a.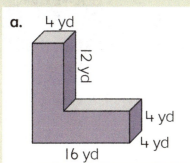

b.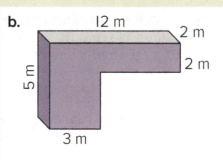

2. Solve each equation. Show each step in your thinking.

a. $v + 18 = 41$

b. $52 = b + 8$

c. $t + 1.4 + 0.5 = 7.5$

d. $d + \frac{5}{12} = 1\frac{8}{12}$

Preparing for Module 8

Write each of these as a multiplication equation. Then complete the equation. Show your thinking.

a. $\frac{3}{4}$ of 9 =

b. $\frac{7}{8}$ of 12 =

c. $\frac{2}{3}$ of 11 =

d. $\frac{3}{5}$ of 18 =

7.11 Algebra: Solving word problems (subtraction and division)

Step In
How would you solve this problem?

Richard borrows a novel from the school library. He starts reading on Saturday and reads 106 pages. He returns to school and notices that he has only 7 pages left to read. How many pages were in the novel?

> I'll let n represent the total number of pages in the novel. That's $n - 106 = 7$.

Dena solves the problem like this.

What steps does she follow?

Why does she add 106 to each side of the equation?

How does each equation remain balanced?

$$n - 106 = 7$$
$$n - 106 + 106 = 7 + 106$$
$$n = 113$$

Marcos has to read a novel for an assignment.

He divides the total number of pages by 9, which is the number of days before the assignment is due. He calculates that he needs to read 13 pages each night to finish the book on time.

How many pages are in the novel?

Let b represent the number of pages.

Write the missing numbers. Then describe the steps that are followed.

$$b \div 9 = 13$$
$$\boxed{} \cdot 9 = \boxed{} \cdot 9$$
$$b = \boxed{}$$

Step Up

1. Solve this problem. Write the missing numbers to show your thinking.

Giselle is working on her college application. She has filled in 3 pages. She notices that there are still 8 more pages to complete. How many pages are in the application? Let a represent the number of pages.

$$a - \boxed{} = \boxed{}$$
$$a - \boxed{} + \boxed{} = \boxed{} + \boxed{}$$
$$a = \boxed{}$$

2. Write an equation to match each problem. Then solve the equation. Remember to include the appropriate unit.

a. A baker makes a loaf of banana bread, splitting it into 11 equal portions that each weigh 2 oz. What was the mass of the whole loaf of bread? Let m represent the mass of the whole loaf.

b. Deana collects sports cards. She trades 15 baseball cards for 20 football cards, and still has 46 baseball cards left. How many baseball cards did she have before the trade? Let b represent the number of cards she had before the trade.

c. Shares in a company drop by $0.37 in one day. The current share price is $1.85. What was the share price at the start of the day? Let s represent the share price at the start of the day.

d. Some oranges are cut into fourths. Andre counts 24 pieces. How many whole oranges were there before? Let z represent the number of oranges.

Step Ahead

Write an equation to represent the problem. Then solve the problem and write the answer below.

It costs $700 to stay 4 nights at a beach resort. It costs $800 to stay 5 nights at a mountain resort. What is the difference in cost between one night's accommodation at each resort? Let c represent the difference in cost.

7.12 Algebra: Solving word problems (all operations)

Step In Read each of these problems.

A Samuel is 60 years old. He says that he is 4 times older than Hailey and 3 times older than Cooper. How old is Cooper? Let g represent the age of Cooper.

B James cuts $3\frac{1}{2}$ feet off a length of rope. The rope is now $2\frac{1}{2}$ feet long. What was the length of the rope before it was cut? Let g represent the length of the rope before it was cut.

C A cashier collects $17.50 from three friends buying meals. The first friend pays $4.50. The second friend pays $5. How much did the third friend pay? Let g represent the amount paid by the third friend.

D Books are being put into a bookcase. 35 books are put on each of the 5 shelves. Some books are thicker than others. How many books were put into the bookcase? Let g represent the total number of books.

How would you solve each problem? What operations would you use?

What equation would you write to match each problem?

Katherine writes these equations.

Which problem is she trying to solve?

What equations should she write next?

$$4.50 + 5 + g = 17.50$$
$$9.50 + g = 17.50$$

She needs to find the value of g. To keep the equation balanced, she has to subtract 9.50 from each side of the equation.

What steps would you follow to solve problem A?

Step Up 1. Solve these problems from the top of the page.

Problem B	Problem D

2. Write an equation to match each problem. Then solve each problem. Remember to include the appropriate unit.

a. Ricardo and Liam are roommates. Ricardo contributes $62.50 toward the phone bill. The total amount due is $103.10, plus an additional $5 late payment fee. What amount does Liam need to pay? Let p represent the amount that Liam pays.

b. Kimie's gym fees are overdue. She pays $50 of her fees on Monday. Her statement says that she now owes $72.50. She hopes to pay another $50 next week. What was the total amount of her gym fees? Let g represent the total amount of her gym fees.

c. Caleb receives a parking ticket. He agrees to make 6 monthly payments of $24.40. What amount was on the parking ticket? Let f represent the amount on the ticket.

d. Gemma paid $12.80 for 4 cups of coffee, then received one more cup at no cost. She split the total cost by 5 to calculate the new cost for each cup of coffee. What is the new cost per cup? Let c represent the new cost per cup.

Step Ahead Write a word problem to match one of these equations. Circle the equation you chose.

$5g = \frac{5}{8}$

$10 + 3.2 + h = 20.1$

$n - 10 = -1$

$b \div 5 = 0.4$

7.12 Maintaining concepts and skills

Think and Solve

A playground was built by combining 2 rectangular spaces of 5 m × 3 m. The overlapped rectangular space measures 2 m × 1 m.

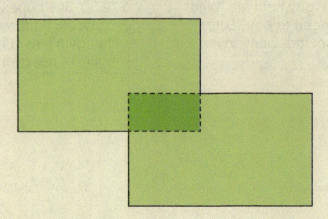

a. What is the perimeter of the whole playground?

b. What is the area of the whole playground?

Words at Work

Write a division word problem that involves a variable. Then show the solution.

Ongoing Practice

1. Solve each problem. Show your thinking.

a. Jessica has two boxes. Box A measures 5 inches by 6 inches by 12 inches. The volume of Box B is half of the volume of Box A. The height of Box B is 10 inches. What could be the other dimensions of Box B?

b. Ryan built a sandbox that measures 5 ft by 3 ft by 1 ft. Sand costs $8.50 per cubic foot. How much sand will Ryan need to fill the sandbox?

2. Solve each equation. Show each step.

a. $5e = 40$

b. $56 = 7x$

c. $21 + 9 = 3y$

d. $5p + p = 4 + 20$

Preparing for Module 8

Write the missing numbers in the diagram. Then complete the equation.

a. $\frac{2}{3} \div \frac{2}{5} =$

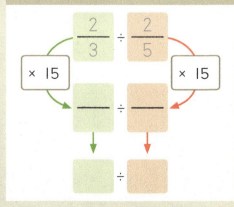

b. $\frac{1}{2} \div \frac{3}{5} =$

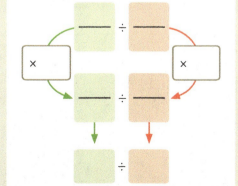

c. $\frac{4}{5} \div \frac{3}{4} =$

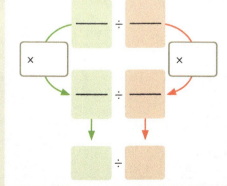

Working Space

8.1 Ratio: Linking part-whole ratios to fractions

Step In

Look at this picture of fruit. What is the relationship between the number of apples and the number of oranges? What is the relationship between the number of apples and the total amount of fruit?

How are the two questions different?

> The first question is comparing the two parts, while the second question is comparing one of the parts with the total.

What are some different ways you could record each answer?
Mary wrote these statements for the second question.

> The ratio of apples to fruit is 4:10.
> The apples are $\frac{4}{10}$ of the total amount of fruit.

> If a ratio describes a **part-whole** relationship, then it can be written as a **common fraction** or **decimal fraction**.

What does the numerator tell you? What does the denominator tell you?
What matching statement involving decimal fractions could you write?

Step Up

1. Shade the ○ beside the type of relationship that exists between the quantities in each word problem. Then record the relationship, using a colon for part-part ratios and a fraction for part-whole ratios.

   a. Max bought 13 pop songs and 8 rock songs. What is the ratio of rock songs to pop songs?
   ○ Part-part
   ○ Part-whole

   b. In a class of 25 students, 15 are boys. What is the ratio of girls to total number of students?
   ○ Part-part
   ○ Part-whole

   c. There are 3 cups of dry ingredients, 2 cups of which are flour. What is the ratio of flour to all dry ingredients?
   ○ Part-part
   ○ Part-whole

   d. 7 eggs have been used out of 12. What is the ratio of remaining eggs to the original total?
   ○ Part-part
   ○ Part-whole

   e. A test was graded on a scale of 1 to 10. Pamela got a 7. What is the ratio of her score compared to the total?
   ○ Part-part
   ○ Part-whole

   f. A bus has 35 passengers. 14 are children and the rest are adults. What is the ratio of children to adults?
   ○ Part-part
   ○ Part-whole

2. A problem involving a part-whole ratio has an answer of 3:7. Write a word problem to match.

3. The ratio of adults to all the people in a room is 3:5.

   a. Complete the table.

Adults	Total	Children
3	5	
6	10	
9	15	
	20	
	25	

   b. How many adults are in a room of 100 people? Show your thinking.

   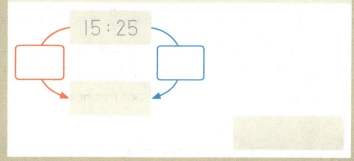

4. Seven-tenths of sixth grade students finished a science project by the due date.

   a. Complete the table.

Finished	Total	Unfinished
7	10	
14	20	
21	30	
	40	
	50	

   b. How many students completed the project if there are 100 students in Grade 6? Show your thinking.

   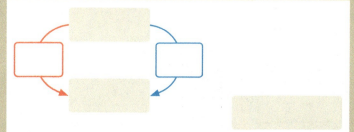

Step Ahead

Abey plants a sunflower seed and Victor plants a corn seed. After 15 days, Abey measures her sunflower and records its growth as the fraction $\frac{16.5}{15}$. After 45 days, Victor measures his corn plant and records its growth as the fraction $\frac{36}{45}$. If the plants continue to grow at the same rates, whose plant is growing faster? Explain your strategy using words and fractions.

8.2 Ratio: Relating fraction representations of ratio

Step In The large square represents one whole. How could you write the amount that is shaded?

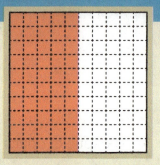

Which of the answers below are correct representations of the shaded part? How do you know?

$\frac{50}{100}$ $\frac{5}{10}$ $\frac{1}{2}$ 0.5   0.50   50:100

Complete this diagram to show how $\frac{50}{100}$ is equivalent to $\frac{5}{10}$ and $\frac{1}{2}$.

What other representations of part-whole ratios could you write to match the shaded part of the square?

How could you calculate the decimal fraction for $\frac{2}{5}$?

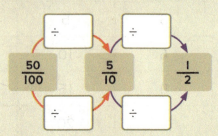

Step Up

1. Each large square is one whole. Write the amount that is shaded as a common fraction, a decimal fraction, and a ratio.

a.
☐ = ☐ = ☐

b.
☐ = ☐ = ☐

c.
☐ = ☐ = ☐

d.
☐ = ☐ = ☐

2. On this number line, the distance between 0 and 1 is one whole. Draw a line to show the position of each common and decimal fraction. Be as accurate as possible.

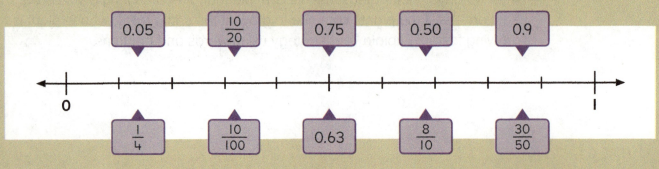

284

3. Complete each equation to show each decimal fraction as a common fraction.

a. $0.1 = \dfrac{\Box}{10}$ b. $0.35 = \dfrac{\Box}{100}$ c. $0.75 = \dfrac{\Box}{100}$ d. $0.85 = \dfrac{\Box}{100}$

4. Complete each equation to show each common fraction as a decimal fraction.

a. $\dfrac{1}{4} = \Box$ b. $\dfrac{3}{5} = \Box$ c. $\dfrac{75}{100} = \Box$ d. $\dfrac{9}{10} = \Box$

5. Circle all the answers that correctly match each statement.

a. A car travels halfway in a trip of 50 miles. Distance traveled out of total trip distance:	**b.** 78 out of 100 students are at lunch. Students at lunch out of total students:
$\dfrac{2}{1}$ $\dfrac{50}{100}$ $50:100$ 0.5	$\dfrac{78}{100}$ $78:100$ $\dfrac{39}{50}$ 7.8
c. 2 seats empty in a row of 10 seats. Ratio of empty seats to all seats:	**d.** $20 spent of $25. Amount spent out of total:
$2:10$ $\dfrac{1}{5}$ 0.4 $\dfrac{8}{10}$	$80:100$ $\dfrac{20}{25}$ $\dfrac{3}{5}$ 0.75
e. 7 eggs remaining in a carton of 10. Ratio of eggs used to original amount:	**f.** 3 in 5 people in the US are younger than 45. People under 45 out of total US population:
$\dfrac{7}{10}$ $\dfrac{30}{100}$ $3:10$ 0.03	$\dfrac{3}{5}$ $5:3$ 0.60 $\dfrac{15}{25}$

Step Ahead

One hundred students vote for their favorite color. 12 vote blue, 18 vote red, 25 vote yellow, 30 vote green, and the remainder vote brown. Color the grid to show the votes, then complete the table to show the amount in different formats.

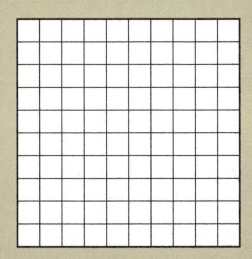

Color	$\dfrac{\Box}{\Box}$	$\Box.\Box\Box$	$\Box:\Box$
blue			
red			
yellow			
green			
brown			

8.2 Maintaining concepts and skills

Computation Practice
Lactuca sativa is the scientific name for which vegetable?

★ Complete the equations. Find each quotient in the grid below and cross out the letter above. Then write the remaining letters at the bottom of the page. Some quotients are used more than once.

524 ÷ 4 =
138 ÷ 6 =
276 ÷ 6 =

183 ÷ 3 =
535 ÷ 5 =
665 ÷ 5 =

192 ÷ 4 =
405 ÷ 9 =
684 ÷ 4 =

360 ÷ 9 =
590 ÷ 5 =
324 ÷ 3 =

156 ÷ 6 =
364 ÷ 4 =
270 ÷ 9 =

840 ÷ 2 =
448 ÷ 8 =
123 ÷ 3 =

516 ÷ 4 =

H	A	L	K	S	E	T	E	R	N
41	91	199	56	118	250	45	131	46	420
W	H	I	T	C	Q	T	C	I	L
107	171	48	303	40	133	159	56	26	118
T	H	D	T	U	Z	U	C	E	V
30	61	108	23	218	40	420	101	659	129

Write the letters in order from the ✱ to the bottom-right corner.

Ongoing Practice

1. Solve each problem. Show your thinking.

a. A bag of popcorn is $\frac{9}{15}$ full. It is shared equally among 3 people. There is no popcorn left in the bag. How much popcorn did each person get?

b. A piece of rope is $\frac{6}{9}$ of a meter long. Janice cuts the rope into shorter pieces that are $\frac{2}{9}$ of a meter long. How many pieces did she cut?

2. The ratio between the sugar content and the total mass of a serving of white chocolate truffles is 9:24.

a. Complete the table.

Sugar (g)	Total (g)	Other (g)
9	24	
18	48	
36	96	
	120	
	168	
	240	

b. How much sugar will be in 360 grams of white chocolate truffles? Show your thinking.

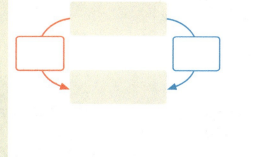

Preparing for Module 9

Write **<** or **>** to make each statement true.

a. 0.74 ◯ 4.7

b. 0.019 ◯ 0.01

c. 0.412 ◯ 0.405

d. 0.08 ◯ 0.88

e. 0.165 ◯ 0.156

f. 0.001 ◯ 0.010

g. 0.606 ◯ 0.676

h. 0.925 ◯ 0.952

i. 0.001 ◯ 0

8.3 Ratio: Introducing percentage (area model)

Step In   This large square represents one whole.

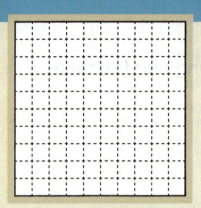

Shade the grid to show $\frac{75}{100}$.

What are some other ways you can write the same amount?

What do you know about percentage?
Where have you heard the term used before?

> **Per** means *for each* and **cent** means *100*. Together, **percent** means *for each hundred*. The symbol for percent is %.

75% of the grid is now shaded.

What percentage of this grid is shaded? How do you know?

Write the missing numbers to show equivalent amounts.

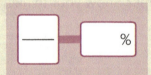

Step Up

1. Shade the grid to show the common fraction. Then write the equivalent percentage.

a. $\frac{45}{100}$ ☐ %

b. $\frac{82}{100}$ ☐ %

c. $\frac{30}{100}$ ☐ %

288

2. Write or shade the equivalent values.

a.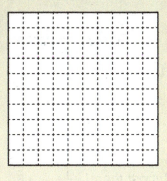

$\dfrac{7}{10}$ $\dfrac{}{100}$ 70%

b.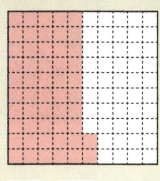

$\dfrac{}{25}$ $\dfrac{52}{100}$ ___ %

c.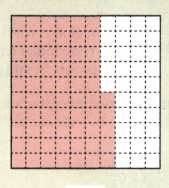

0.65 $\dfrac{}{100}$ ___ %

d.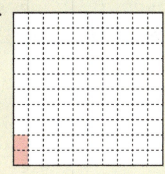

___ : ___ $\dfrac{2}{100}$ ___ %

e.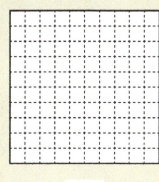

___ $\dfrac{43}{100}$ 43%

f.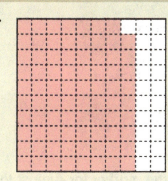

___ : ___ $\dfrac{79}{100}$ ___ %

g.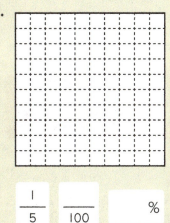

$\dfrac{1}{5}$ $\dfrac{}{100}$ ___ %

h.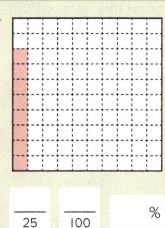

$\dfrac{}{25}$ $\dfrac{}{100}$ ___ %

i.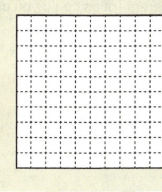

___ : ___   90 : 100   ___

Step Ahead   Think about a grid to help you solve this problem.

A radio station asked their listeners to vote for the type of music they liked best. The survey found that 45% of listeners like rock, 30% like pop, and the rest like hip-hop. What percentage of listeners like hip-hop?

8.4 Ratio: Consolidating percentage (number line model)

Step In

A club is raising funds for a charity. They make a picture to track their progress and mark major milestones.

Calculate the fraction of the total amount to be raised at each milestone. Then determine the equivalent percentage.

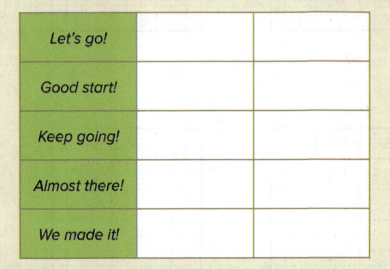
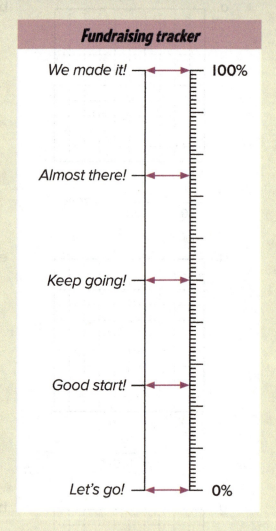

Look at the marker for *Almost there!*

What is the ratio of funds raised to the goal amount?

What different ratios could you use to express this relationship?

Complete the missing numbers in this diagram.

Step Up

1. Write the common fraction and percentage shown by each pair of arrows.

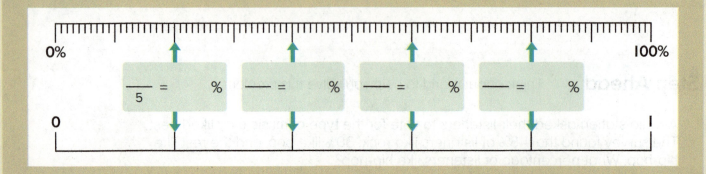

2. Complete these to show equivalent common fractions and percentages.

a.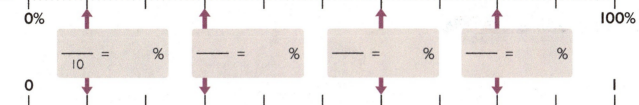

$\frac{\square}{10} = \square\%$ $\frac{\square}{\square} = \square\%$ $\frac{\square}{\square} = \square\%$ $\frac{\square}{\square} = \square\%$

b.

$\frac{\square}{\square} = \square\%$ $\frac{\square}{\square} = \square\%$ $\frac{\square}{\square} = \square\%$

3. Write the missing numbers. All the ratios are part-whole ratios.

a. $\frac{2}{5} = \frac{\square}{100} = \square\%$

b. $19\% = \frac{\square}{\square} = \square .$

c. $3:5 = \frac{\square}{100} = \square .$

d. $\frac{\square}{10} = \frac{20}{100} = \square\%$

e. $\frac{4}{5} = \square : 100 = \square\%$

f. $\frac{7}{10} = \square\% = \square : \square$

g. $\frac{4}{20} = \frac{\square}{100} = \square\%$

h. $\square\% = 0.35 = \square : \square$

i. $98\% = \square : \square = \frac{\square}{\square}$

j. $\frac{1}{4} = \frac{\square}{100} = \square\%$

k. $\frac{\square}{\square} = \frac{30}{100} = \square : \square$

l. $\frac{5}{100} = \square\% = \frac{1}{\square}$

Step Ahead

Students were asked to vote for their favorite movie type. The survey found that three-fifths of students enjoy funny movies, two-tenths like scary movies, and the rest prefer action movies.

a. What percentage of students liked action movies? $\square$ %

b. 200 students participated in the survey.
Write the number of students who voted for each type of movie.

$\square$ funny $\square$ scary $\square$ action

8.4 Maintaining concepts and skills

Think and Solve

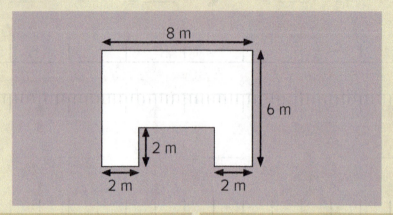

a. What is the perimeter of this shape? ☐

b. What is the area of this shape? ☐

Words at Work Research and write about how percentage is used in everyday life.

Ongoing Practice

1. Solve each problem. Show your thinking.

a. $12 \div \frac{2}{5} = \square$

b. $15 \div \frac{3}{4} = \square$

c. $20 \div \frac{5}{9} = \square$

d. $18 \div \frac{2}{8} = \square$

2. Circle all the answers that correctly match the context.

a. 8 roses in every bunch of 24 flowers. Number of roses out of a bunch of flowers:

$3:1$ $\frac{8}{24}$ $16:48$ $\frac{1}{3}$

b. Out of 100 students, 42 bike to school. Students who bike out of total students:

$\frac{42}{100}$ $42:100$ $\frac{21}{50}$ 4.2

c. 5 cans of cola for every 10 cans sold. Ratio of cans of cola to cans sold:

$5:10$ $\frac{1}{2}$ 0.5 $10:5$

d. $5 donated to charity out of $50 earned. Amount donated out of total amount earned:

$5:50$ 0.5 $10:100$ $\frac{1}{10}$

e. 5 packs remaining in a box of 20 packs. Ratio of packs used to original amount:

$\frac{1}{4}$ $\frac{75}{100}$ $5:20$ 0.75

f. 4 in 5 people own a car. People who own cars out of total people:

$\frac{4}{5}$ $5:4$ 0.8 $\frac{20}{25}$

Preparing for Module 9

Circle any 4 integers. Then draw a number line to show their positions.

0   −9   5   −10   27   15

8.5 Ratio: Simple percentages of quantities

Step In

Layla is saving money to buy this new pair of skates. After a few weeks she has saved 25% of the total amount. How much money has she saved? How do you know?

I know that 25% is equivalent to one-fourth. That's the same as dividing by 4.

A few weeks later, Layla has saved 75% of the total cost.

How much total money has she saved so far? How do you know?

Well, 75% is equivalent to three-fourths. I can find one-fourth of $80 and multiply that by 3.

Nancy is also saving for the same pair of skates. She has saved 20% of the total cost. How can you calculate the amount of money she has saved?

Cody drew a **double number line** to calculate the answer.

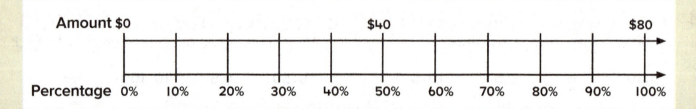

What information does the double number line tell you?
What steps should Cody take next to calculate the answer?
What information does he need to include on the number line?

Step Up

1. Use the double number line to determine the percentage.

20% of 86 lb = ☐

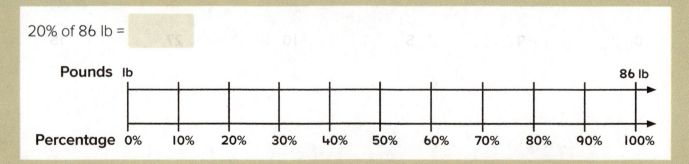

2. Use the double number lines to determine each percentage. Be sure to label each number line appropriately and include the appropriate unit in your answers.

a. 25% of 76 ft = ☐

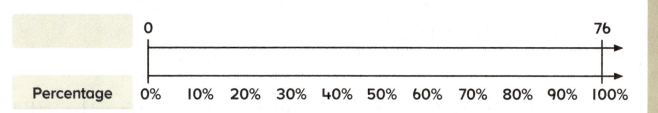

b. 75% of $90 = ☐

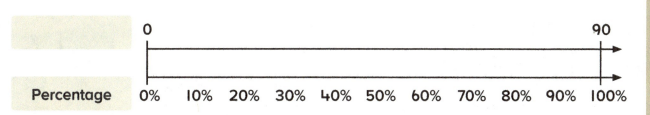

3. Write equations to calculate each percentage. Remember to include the appropriate units in your answer.

a. 50% of 54.25 kg = ☐

b. 25% of $93 = ☐

c. 20% of 65 yd = ☐

d. 75% of 124 oz = ☐

Step Ahead

Some unit fractions are equivalent to whole number percentages. For example, $\frac{1}{2}$ is equivalent to 50%. Some unit fractions are not equivalent to whole number percentages. For example, $\frac{1}{16}$ is equivalent to 6.25%.

Identify two unit fractions not shown on this page that are equivalent to whole number percentages and two that are not. Write them below. Show your thinking on page 318.

─── = ☐   ─── = ☐   ─── = ☐   ─── = ☐

8.6 Ratio: Percentages of collections

Step In

Teresa wants to save $60. She has saved 20% of the total. How much has she saved?

Charlie drew a double number line to figure it out. What has he shown so far? What should he do next?

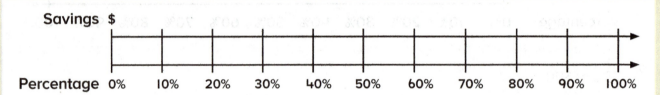

Mako used the following method.

$$\frac{20}{100} \cdot \frac{60}{1} = \frac{1200}{100} = \frac{12}{1} = \$12$$

What process does she follow? How else could you find the solution?

> I know that 10% of $60 is $6. Double that is $12.

John buys a book online for a total cost of $40. Of the total cost, 15% is for shipping. How much does he pay for shipping?

> I know that 10% of $40 is $4. Half of that is $2. If I add the two amounts together I can find 15% of $40.

Step Up

1. Use the double number line to calculate the percentage of each amount. Remember to include the appropriate unit in your answer.

a. 30% of $70 =

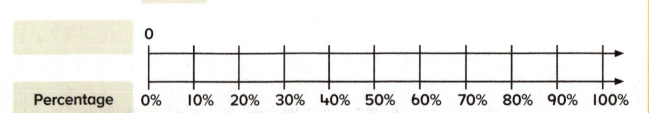

b. 60% of 120 mi =

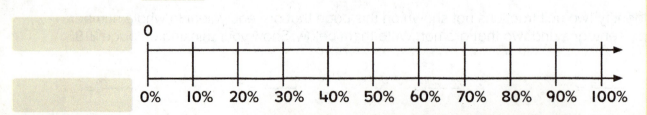

2. Calculate each percentage. Show your thinking. Remember to include the appropriate units in your answer.

a. 20% of $150 =

b. 15% of 60 km =

c. 90% of 90 lb =

d. 35% of 50 qt =

3. Solve each problem. Show your thinking. Remember to include the appropriate units in your answer.

a. Thomas usually spends 40% of homework time on math. If he does an hour of homework, how much time does he spend on math?

b. 70% of 250 students live in the same part of town. 15% of them ride the bus to school. How many students come from the same neighborhood?

c. In a factory, 25% of every 1,500 new cars made are gray. How many cars are not gray?

d. Mia earns $50,000 per year and spends about 10% of it on food and 30% on rent. How much rent does she pay each year?

Step Ahead

Finding a percentage of a quantity is the same as multiplying with that percentage. For example, 25% of 90 is equivalent to 25% • 90 or 90 • 25%. Complete the number trail.

8,000 → ×20% → 1,600 → ×50% → ☐ → ×60% → ☐ → ×10% → ☐ → ×25% → ☐ → ×75% → ☐

8.6 Maintaining concepts and skills

Computation Practice

★ Complete each equation. Then write each letter above its matching answer at the bottom of the page to discover an interesting animal fact. Some letters are used more than once.

1.5 + 7.8 =	p	3.1 + 0.46 =	o	3.6 − 2.5 =	f
2.16 − 0.25 =	t	3.5 + 11.8 =	c	5.01 − 3.8 =	r
5.7 + 4.09 =	a	1.45 + 0.35 =	p	8.01 − 5.9 =	h
2.15 − 0.6 =	i	4.8 − 0.9 =	e	2.4 − 0.25 =	a
5.5 + 2.61 =	m	0.65 + 0.6 =	n	1.02 + 0.65 =	s
1.4 + 0.55 =	t	3.05 − 0.44 =	u	0.25 + 2.56 =	r
2.6 + 0.23 =	h	4.03 + 0.24 =	n	4.5 − 1.08 =	s
3.5 − 2.8 =	a				

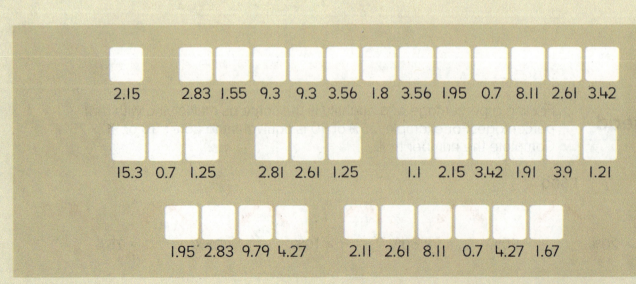

2.15   2.83 1.55 9.3 9.3 3.56 1.8 3.56 1.95 0.7 8.11 2.61 3.42

15.3 0.7 1.25   2.81 2.61 1.25   1.1 2.15 3.42 1.91 3.9 1.21

1.95 2.83 9.79 4.27   2.11 2.61 8.11 0.7 4.27 1.67

Ongoing Practice

1. Complete each equation. Show your thinking.

a. $5 \div \frac{2}{3} =$ ☐

b. $12 \div \frac{5}{6} =$ ☐

c. $9 \div \frac{4}{5} =$ ☐

d. $17 \div \frac{3}{8} =$ ☐

2. Write or shade equivalent values.

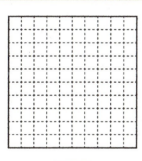

$0.\_\_\_ = \frac{45}{100} = \_\_\_ \%$

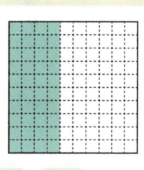

$\frac{\_\_\_}{100} = \_\_\_ \%$

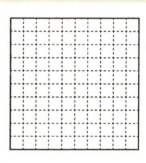

$\frac{35}{100} = \_\_\_ \% = \_\_\_$

Preparing for Module 9

Look at the dot plot and answer the questions.

Grade 6 Reading

Number of pages read in 30 minutes

a. How many students took part in reading? ☐

b. How many students read exactly 8 pages in 30 minutes? ☐

c. How many students read more than 8 pages in 30 minutes? ☐

d. How many students read fewer than 8 pages in 30 minutes? ☐

8.7 Ratio: Percentages of numbers less than ten

Step In Leila is calculating 30% of $5.

She knows that 30% of each $1.00 is 30¢, so she reasoned that 5 • 30¢ = 150¢ = $1.50.

Aaron wrote this equation to calculate the answer.

Why could Aaron use $\frac{3}{10}$ instead of $\frac{30}{100}$?
What other method could you use to calculate 30% of $5?
In which ways could you report the answer if the unit was meters and not dollars?

30% of $5 = $\frac{3}{10}$ of 5 or $\frac{3}{10} \cdot \frac{5}{1} = \frac{15}{10} = \$1.50$

Step Up

1. Use Aaron's method to calculate each percentage. Be sure to include the appropriate units in your answer.

a. 20% of 4 mi	b. 10% of 3 kg	c. 30% of 2 lb
d. 50% of $2.50	e. 20% of $1.50	f. 80% of 5 lb

2. Calculate each percentage so the answer is a whole number. You will need to convert to an appropriate unit.

> 1 m is equivalent to 100 cm.
> 1 km is equivalent to 1,000 m.

a. 20% of 3 m =

b. 55% of 5 km =

3. Estimate who has saved more money. Circle their name. Then calculate how much that person saved. Show your thinking.

Damon saved 30% of $4, while Ruby saved 20% of $5.

4. Calculate the answers. Show your thinking.

a. A survey found that 20% of people in 5 states earn more than $900 per week. What fraction of people surveyed earn less than $900 per week?

b. 75% of every 4 children get an allowance. What fraction of every 1,000 children do not get an allowance?

Step Ahead

A factory recycles 15% of its waste products. Over one 12-hour shift, the machines records 25 lb of waste every 1.5 hours. How many pounds of waste are recycled that day?

8.8 Ratio: Using unit percentages

Step In

8% of the students at Clever Clogs Elementary School play chess on Monday. If there are 200 students at the school, how many play chess on Monday?

Peta thought about it this way. $\frac{8}{100} \cdot \frac{200}{1} = \frac{1600}{100} = \frac{16}{1} = 16$

How else could you find the answer?

"One-tenth of 200 is 20 and one-tenth of 20 is 2. That tells me what 1% is. If 1% of 200 is 2, then 8% is 8 • 2 = 16."

14% of the students play basketball on Tuesday afternoon. If there are 200 students how many play basketball? How do you know?

Step Up

1. Calculate each percentage. Show your thinking.

    a. 12% of 300 =

    b. 12% of 30 =

    c. 12% of 3 =

2. Describe what you notice about the answers for Question 1.

3. Calculate each percentage. Show your thinking.

    a. 9% of 160 =

    b. 22% of 40 =

    c. 33% of 180 =

4. Solve each problem. Show your thinking.

   a. Nathan wants to buy a game for $80. He has saved 17% of the total cost. How much money has he saved?

   b. A game has a file size of 1,300 megabytes. Laura has downloaded 7% of the file. How many megabytes has she downloaded?

   c. There were approximately 7,500,000,000 people on earth in 2016. It is estimated that 1% of the world's population own nearly half of the global wealth. How many people are in this category?

5. 1,200 people vote for their favorite artist from this list. The results are recorded in the table. Complete the missing parts of the table. Show your thinking.

Favorite Artist		
Artist	Percentage of votes	Total
Vincent van Gogh	24%	
Pablo Picasso	19%	
Claude Monet	16%	
Andy Warhol	14%	
Leonardo da Vinci	13%	
Michelangelo	8%	
Unsure	6%	

Step Ahead

Oscar knows some benchmark fractions as percentages. For example, $\frac{1}{4}$ is 25%, $\frac{1}{5}$ is 20%, and $\frac{1}{2}$ is 50%. What is $\frac{1}{3}$ as a percentage? Explain why.

8.8 Maintaining concepts and skills

Think and Solve

Imagine you cut this prism into 2 smaller prisms so that one has double the **volume** of the other.

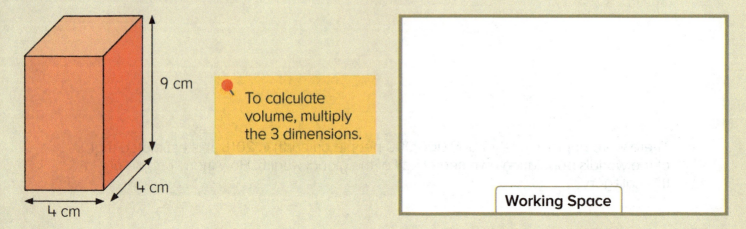

To calculate volume, multiply the 3 dimensions.

Working Space

Write the length, width, height, and volume of each smaller prism.

Words at Work

Imagine your friend was away from school when you learned about percentages. Explain in words the steps you would use to determine 25% of $120.

Ongoing Practice

1. Write an equation to match each problem. Let *x* stand for the unknown value.

a. The area of Carlos' courtyard is 54 ft². His courtyard is larger than Sheree's by 5.5 ft². What is the area of Sheree's courtyard?

b. Mr. Jones ordered new school books for his 3 children. Each child needs 3 large books, 8 small books, and a math book. How many books did he order in total?

FROM 6.4.2

2. Complete these to show equivalent values.

FROM 6.8.4

a. $16\% = \dfrac{\boxed{}}{100} = \boxed{}$

b. $\dfrac{3}{5} = \dfrac{\boxed{}}{100} = \boxed{} : 5$

c. $\boxed{}\% = \boxed{} : 100 = 0.29$

d. $\boxed{}\% = \boxed{} : 10 = \dfrac{8}{10}$

e. $\boxed{}\% = \boxed{} : 100 = 0.87$

f. $\dfrac{\boxed{}}{25} = \dfrac{\boxed{}}{100} = \boxed{} : 25$

g. $\dfrac{2}{25} = \dfrac{\boxed{}}{100} = 2 : \boxed{}$

h. $\boxed{}\% = 30 : \boxed{} = \dfrac{3}{10}$

i. $60\% = 3 : \boxed{} = \boxed{}$

j. $\boxed{}\% = \boxed{} : 20 = 0.45$

k. $\dfrac{9}{50} = \dfrac{\boxed{}}{\boxed{}} = 9 : \boxed{}$

l. $\boxed{}\% = 3 : \boxed{} = \dfrac{3}{4}$

Preparing for Module 9

Each small square represents 1 m². Draw a rectangle around each triangle to help you calculate the area of each triangle.

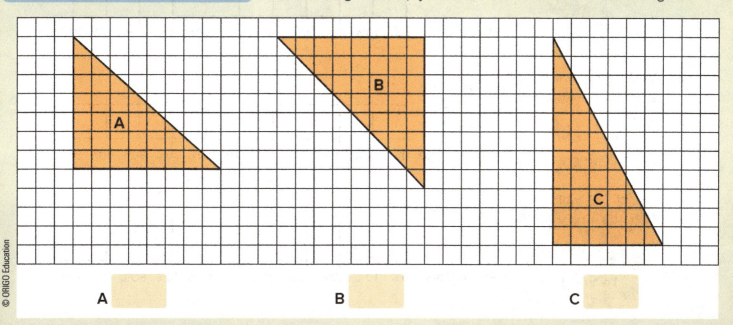

A ⬚ B ⬚ C ⬚

ORIGO Stepping Stones · Grade 6 · 8.8

8.9 Ratio: Finding the whole given a part and the percentage

Step In

Nicole saves 30% of her allowance each week. This week she saved $6. How much allowance did she get this week?

Stella drew a double number line. How can she use the diagram to find the answer?

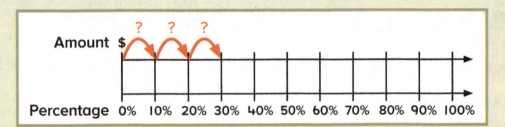

Dixon drew a relationship diagram to show his thinking. Write the missing numbers in the spaces.

What is the same about these methods?

Which method would you use to calculate the original price if $21 is 25% of the original price? Is there another method you could use instead?

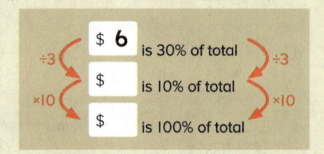

$6 is 30% of total
$ ___ is 10% of total
$ ___ is 100% of total

Step Up

1. Complete each calculation. Show your thinking on the double number lines.

a. I spent $36, which is 40% of what I earned.
   I earned ___.

b. I spent $24, which is 60% of what I earned.
   I earned ___.

c. I spent $12, which is 20% of what I earned.
   I earned ___.

2. Each club has monthly fundraising goals. They record the current amount raised as a percentage of the goal amount. Draw a relationship diagram to establish the goal amount for each club.

a. $120 raised which is 3% of goal

b. $480 raised which is 12% of goal

3. Solve each problem. Show your thinking. Be sure to include the appropriate units in your answer.

a. Gavin is trying to set a record for jump rope. He has jumped rope for 105 minutes, which is 5% of his goal. What is his goal in hours?

b. Kinu bought tickets to a concert. Her mom contributed $84 toward the tickets which was 60% of the total price. What was the full price of the tickets?

c. A school wants to build a beach volleyball court. The PTA will pay for 40% of the sand required. If the PTA pays for 72 tons of sand, how many total tons of sand are needed?

d. A warehouse has 50 equally stacked shelves. Daniela counts 25% of the boxes on one shelf and finds there are 30 boxes. How many boxes are in the warehouse?

Step Ahead

William paid $40 for an item after he received a 20% discount. William's friend stated that the original price must have been $48. Is his friend correct? Show your thinking. Use page 318 if you need more space.

8.10 Division: Introducing the invert-and-multiply method (common fractions)

Step In David noticed something interesting about division. Look at his thinking on the right.

$$12 \div 3 = 12 \cdot \frac{1}{3}$$

so

$$\frac{12}{1} \div \frac{3}{1} = \frac{12}{1} \cdot \frac{1}{3}$$

How are the two equations similar?
How are the divisor and second factor related?

David then tried a slightly different problem.
Why does his method also work this time?

$$4 \div \frac{1}{5} = 4 \cdot \frac{5}{1}$$

so

$$\frac{4}{1} \div \frac{1}{5} = \frac{4}{1} \cdot \frac{5}{1}$$

I know that 1 divided by $\frac{1}{5}$ is 5, so 4 divided by $\frac{1}{5}$ is 4 times as much as 5. That's the same as 4 · 5.

If dividing by a whole number is the same as multiplying by a unit fraction then the reverse is true as well.

Is this equation true? How do you know?

$$\frac{6}{8} \div \frac{2}{8} = \frac{6}{8} \cdot \frac{8}{2}$$

$$\frac{2}{3} \div \frac{7}{8} = \boxed{} \cdot \boxed{}$$

$$\boxed{}$$

How could you use the same thinking to solve this problem?

Write a multiplication expression to help solve this problem.
Then write the quotient.

If you need to divide one common fraction by another, you can invert the divisor then multiply by the fraction that is produced.
For example, $\frac{4}{7} \div \frac{5}{6} = \frac{4}{7} \cdot \frac{6}{5}$.

To **invert** a fraction means to exchange the values of the numerator and denominator.

Step Up

1. Use the invert-and-multiply method to write equivalent expressions. You do not have to write the quotient.

a. $\frac{3}{5} \div \frac{2}{7}$

b. $\frac{6}{9} \div \frac{3}{4}$

c. $\frac{3}{5} \div \frac{7}{8}$

d. $\frac{9}{12} \div \frac{1}{5}$

308

2. Invert and multiply to solve each division problem. Then complete the equation.

a. $\frac{3}{8} \div \frac{5}{6} = \square$

$\frac{3}{8} \cdot \frac{6}{5} = \square$

b. $\frac{5}{7} \div \frac{4}{9} = \square$

$\square \cdot \square = \square$

c. $\frac{3}{4} \div \frac{7}{8} = \square$

$\square \cdot \square = \square$

3. Solve each problem. Show your thinking and include the appropriate unit in your answer.

a. A piece of yarn is five-sixths of a yard long. Trina starts to cut lengths from it that are one-eighth of a yard long. How many whole lengths can she cut?

b. Underground power is being run along the main street, which is four-fifths of a mile long. A power box must be installed every two-twelfths of a mile. How many power boxes are installed?

c. Felipe walks three and one-fourth miles to the beach and stops every two-thirds of a mile for a drink. How many times does he stop?

d. Four-fifths of one stick of butter is left. Each batch of muffins uses one-third of a stick of butter. How many batches and partial batches of muffins can be made?

Step Ahead

Circle the equation that you would solve using the invert-and-multiply method. Complete the equation. Then write in words why you chose the equation.

$\frac{6}{8} \div \frac{2}{8} = \square$

$\frac{3}{5} \div \frac{2}{3} = \square$

$\frac{3}{4} \div \frac{3}{12} = \square$

ORIGO Stepping Stones • Grade 6 • 8.10

8.10 Maintaining concepts and skills

Computation Practice — What does the word *karate* literally mean?

★ Calculate the answers. Find each answer in the grid below and cross out the letter above. Then write the remaining letters at the bottom of the page.

50% • 340 = 170	25% • 560 = 140	50% • 132 = 66
25% • 144 = 36	50% • 314 = 157	25% • 316 = 79
50% • 276 = 138	25% • 276 = 69	50% • 494 = 247
25% • 976 = 244	50% • 782 = 391	25% • 388 = 97
50% • 590 = 295	25% • 496 = 124	50% • 618 = 309
25% • 696 = 174	50% • 814 = 407	25% • 92 = 23
50% • 902 = 451	25% • 844 = 211	50% • 38 = 19
25% • 68 = 17	50% • 370 = 185	25% • 380 = 95

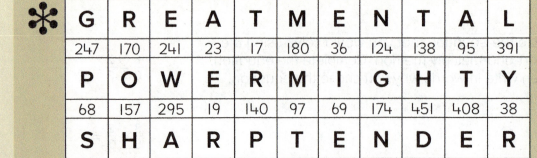

✱	G	R	E	A	T	M	E	N	T	A	L
	247	170	241	23	17	180	36	124	138	95	391
	P	**O**	**W**	**E**	**R**	**M**	**I**	**G**	**H**	**T**	**Y**
	68	157	295	19	140	97	69	174	451	408	38
	S	**H**	**A**	**R**	**P**	**T**	**E**	**N**	**D**	**E**	**R**
	79	96	395	211	244	407	185	27	123	66	309

Write the letters in order from the ✱ to the bottom-right corner.

EMPTY **HAND**

Ongoing Practice

1. Evaluate each expression. Then complete each pair of equations. Show your thinking.

a. 2.56 + 4b   b = 3
☐ = ☐
2.56 + 4b   b = 18
☐ = ☐

b. 24f − 51   f = 5
☐ = ☐
24f − 51   f = 3
☐ = ☐

c. 45 ÷ 5(x)   x = 6
☐ = ☐
45 ÷ 5(x)   x = 10
☐ = ☐

2. Write equations to calculate each percentage. Remember to include the appropriate units in your answer.

a. 20% of 85 mL = ☐

b. 10% of $63 = ☐

c. 25% of 24 yd = ☐

d. 60% of 50 oz = ☐

Preparing for Module 9

Calculate the area of each shape below. Remember to record the correct units.

a.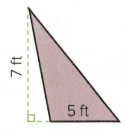

b.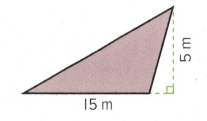

8.11 Division: Consolidating the invert-and-multiply method (common fractions)

Step In

Think about the relationship between each dividend and divisor.
Estimate which expressions will record a quotient that is less than one.

A $\frac{3}{4} \div \frac{3}{8}$
B $\frac{3}{4} \div \frac{8}{10}$
C $\frac{2}{3} \div \frac{8}{5}$
D $\frac{7}{6} \div \frac{4}{5}$

How did you decide?

What strategy would you use to complete each expression?

Solve problems A and C in the space below. Show your thinking.

$\frac{3}{4} \div \frac{3}{8} =$ ☐

$\frac{2}{3} \div \frac{8}{5} =$ ☐

What steps would you follow to solve problems B and D?

> Depending on the denominators, I can calculate the quotient without needing to invert and multiply.

Step Up

1. Use the invert-and-multiply method to solve each division problem. Then complete the equation.

a. $\frac{3}{5} \div \frac{2}{4} =$ ☐

$\frac{3}{5} \cdot \frac{4}{2}$

b. $\frac{2}{6} \div \frac{5}{7} =$ ☐

c. $\frac{3}{9} \div \frac{2}{3} =$ ☐

d. $\frac{4}{7} \div \frac{2}{3} =$ ☐

e. $\frac{2}{5} \div \frac{8}{7} =$ ☐

f. $\frac{10}{6} \div \frac{5}{9} =$ ☐

2. Think carefully about how you would solve each of these. Shade four expressions you would solve using the invert-and-multiply method. Then calculate each quotient in the spaces below.

$\frac{5}{6} \div \frac{2}{12}$ $\frac{1}{3} \div \frac{4}{7}$ $\frac{3}{5} \div \frac{2}{3}$ $\frac{7}{9} \div \frac{4}{5}$

$\frac{9}{15} \div \frac{3}{15}$ $\frac{2}{3} \div \frac{4}{12}$ $\frac{6}{8} \div \frac{2}{8}$ $\frac{4}{9} \div \frac{4}{6}$

a.

b.

c.

d.

3. Solve the remaining problems in Question 2. Show your thinking.

a.

b.

c.

d.

4. How did you decide which problems to shade in Question 2?

Step Ahead Samantha writes a division expression using two of these numbers and she records a quotient of three-fourths. Circle the two numbers that she worked with. Show your thinking.

$\frac{9}{7}$

$\frac{11}{3}$

$2\frac{3}{4}$

ORIGO Stepping Stones • Grade 6 • 8.11

8.12 Division: Consolidating strategies (common fractions)

Step In

Which equations have a quotient greater than one? How do you know?

How would you solve each equation?

Would you use the same strategy for each equation?

What different strategies could you use?

$5 \div \frac{1}{4} =$ $\frac{9}{12} \div \frac{3}{12} =$

$\frac{6}{12} \div 3 =$ $\frac{4}{5} \div \frac{1}{3} =$

$\frac{2}{3} \div \frac{6}{9} =$ $\frac{5}{6} \div \frac{8}{7} =$

I look at the relationship between the dividend and the divisor to decide what strategy to use.

$\frac{2}{3}$ and $\frac{6}{9}$ are equivalent fractions. That's like dividing a number by itself.

For which equations did you invert and multiply?

How did you decide when to use this method?

Step Up

1. Complete each equation. Show your thinking on page 318. Then explain in words the strategy you used to find the quotient.

a. $4 \div \frac{1}{3} = \boxed{12}$

b. $\frac{3}{7} \div \frac{9}{5} =$

c. $\frac{24}{30} \div \frac{8}{30} =$

d. $\frac{9}{6} \div \frac{3}{5} =$

2. Solve each problem. Show your thinking.

a. Jamal has an eight-page assignment to complete. He decides to complete two-fifths of one page each night. How many nights will it take to complete the assignment?

b. Part of a road measuring one and a half miles is being repaired. The repairs are shared equally among six workers. What fraction of the road is each worker expected to repair?

c. Jude is watering his garden. His bucket holds two and a half gallons of water. Each plant is given one-third of a gallon. How many plants is he able to water with one full bucket?

d. A large wheel of cheese weighing four and two-third pounds is cut into smaller blocks. Each small block weighs about half a pound. How many small blocks were cut?

e. All the sandwiches are cut into fourths and put on a tray. One-half of the sandwiches have peanut butter and jelly. One-fourth of the sandwiches have only peanut butter, and the rest have only jelly. Allison counts 15 fourths that have some peanut butter. How many whole sandwiches were made?

Step Ahead

Write numbers to make this equation true. Then write a word problem to match.

$\square \div \square = \frac{2}{5}$

8.12 Maintaining concepts and skills

Think and Solve

This large equilateral triangle is made with 4 smaller congruent equilateral triangles. The ratio of the area of Triangle A to the large triangle can be written as 1:4.

a. What is the ratio of the area of the shaded shape to the large triangle? ☐:☐

b. What is the ratio of the area of Triangle A to the shaded shape? ☐:☐

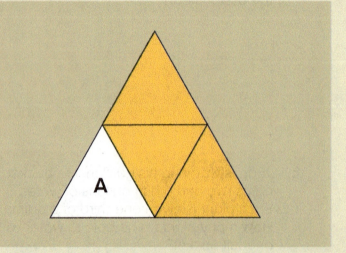

Words at Work

Choose and write words from the list to complete true statements.

a. If a ratio describes a ☐-☐ relationship it can be written as a common or decimal fraction.

b. The ratio 50:100 is ☐ to $\frac{50}{100}$.

c. 75% is ☐ to $\frac{80}{100}$.

d. A double number line can be used to determine a ☐.

e. To calculate 25% of an amount, you ☐ the amount by 4.

f. If you know the value of 40% then you can ☐ the value of 100%.

g. When dividing two common fractions, you can invert the ☐ and change the operation to ☐.

h. The answer in a division problem is called the ☐.

- calculate
- percentage
- multiplication
- quotient
- divisor
- not equal
- part
- whole
- divide
- equivalent

316

Ongoing Practice

1. Circle one number to represent the variable. Then complete the equation. Show your thinking.

a. 2   5   3   9

$3v(4+5)$ $v=$ ▢

▢ = ▢

b. 12   4   3   9

$3^4 \div y$ $y=$ ▢

▢ = ▢

c. 15   5   3   2

$45 \cdot 0.2 \div d$ $d=$ ▢

▢ = ▢

2. Calculate each percentage. Show your thinking and include the appropriate unit in your answer.

a. 75% of $150 = ▢

b. 40% of 70 km = ▢

c. 85% of 120 lb = ▢

d. 35% of 80 gal = ▢

Preparing for Module 9

Draw extra lines if needed to adjust each shape below. Then calculate the area.

a.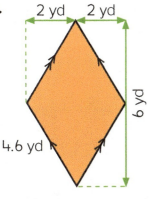
2 yd   2 yd
6 yd
4.6 yd

b.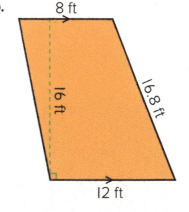
8 ft
16 ft
16.8 ft
12 ft

Working Space

9.1 Algebra: Reviewing inequalities

Step In What do you know about this pan balance?

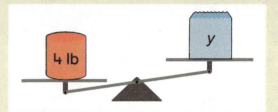

What are the possible values for y?

Complete each inequality below to match the pan balance.

What do you know about the items on the pan balance below?

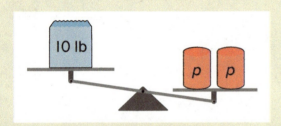

> An **inequality** is a relationship between two quantities that are not equal.

What inequality could you write?
Can you simplify the inequality?

What do you know about the value of $1p$?

> Like equations, inequalities can be simplified by using a common factor. If $10 < 2p$, then $5 < 1p$.

Step Up 1. Complete each inequality to show the possible values for the variable.

a.
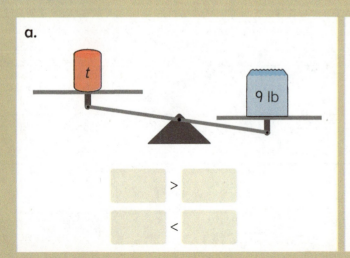

b.
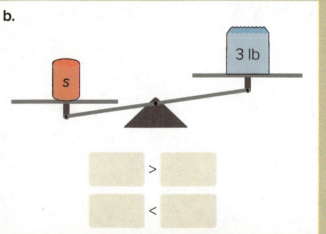

2. Complete each inequality to match each pan balance.

a.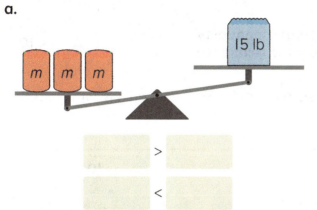

☐ > ☐

☐ < ☐

b.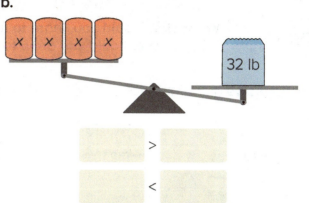

☐ > ☐

☐ < ☐

3. Write **<** or **>** to make each inequality true. You do not need to calculate each expression exactly.

a. $(18 + 4) \div 2$ ◯ $18 \div 2$

b. $16 \cdot 2 \div 4$ ◯ $60 \div 10$

c. $9(5.2 + 2.7)$ ◯ $10 \cdot 8$

d. $10^2 \cdot 0.46$ ◯ $0.46 \div 10$

e. $34 \div 0.2$ ◯ $3.4 \div 0.5$

f. $1.6 \div 0.2$ ◯ $0.2 \div 1.6$

g. $6.24 \div 0.12$ ◯ $6.24 \div 0.06$

h. $0.768 \div 0.7$ ◯ $0.768 \div 0.76$

Step Ahead

Look at the statement $n \div 5$ ◯ $5 \div n$. Which symbol (**<**, **>**, or **=**) can you put in the circle to make the statement true? Will it always be true for any value of n? Explain your thinking.

9.2 Algebra: Showing inequalities on a number line

Step In What story could you share to match this inequality?
What picture could you draw to show the relationship?

$h < 10$

Sofia draws a pan balance.

Paul draws a number line then marks it with color.

How does each picture model the inequality?
What are some of the possible values for h that make the inequality true?

> I think Paul has circled the 10 to show that 10 is not included in the list of possible values. Everything to the left of 10 must be what is possible.

Mark this number line to show this inequality.

$b > 16$

Write an inequality to match the number line below.
Let k represent the variable.

Step Up 1. Show each inequality on the number line.

a. $t < 17$

b. $w > 6$

c. $32 > j$

2. Write an inequality to match each number line. Let *n* represent the variable.

a.

b.

c.

d.

e.

f.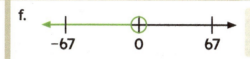

3. Read each statement. Then draw a number line using two different colors to show each inequality.

a. The value of a number is less than 43.

b. The value of a number is greater than −8.

c. The temperature is less than −2°F.

d. Nadia was on the phone for more than $\frac{3}{4}$ of an hour.

Step Ahead Compare the two inequalities. Then write what is the same and what is different between them. $4.2 < 7$ $b < 7$

9.2 Maintaining concepts and skills

Computation Practice

★ Complete the equations. Then write each letter above its matching product at the bottom of the page to discover a fact about the natural world. Some letters are used more than once.

265 • 2 =	u		4 • 125 =	e		5 • 145 =	h
3 • 155 =	d		3 • 225 =	s		3 • 135 =	f
265 • 3 =	s		5 • 115 =	i		2 • 306 =	l
220 • 5 =	i		185 • 5 =	e		4 • 325 =	a
2 • 295 =	a		165 • 4 =	a		5 • 202 =	t
8 • 115 =	u		3 • 245 =	c		9 • 115 =	a
5 • 205 =	p		165 • 6 =	s		5 • 108 =	r
325 • 3 =	o		4 • 105 =	o		2 • 420 =	g
175 • 4 =	m		5 • 201 =	o		5 • 306 =	t
315 • 3 =	o		4 • 124 =	p		4 • 310 =	b

☐ ☐ ☐ ☐ ☐ ☐ ☐ ☐
1,300 840 540 1,005 530 1,025   1,005 405

☐ ☐ ☐ ☐ ☐ ☐ ☐ ☐ ☐ ☐ ☐ ☐ ☐
725 1,100 496 496 420 496 975 1,530 590 700 920 795 925 675

☐ ☐   ☐ ☐ ☐ ☐ ☐   ☐
575 990   735 660 612 612 500 465   1,300

☐ ☐ ☐ ☐ ☐
1,240 612 945 1,035 1,010

◆ 324

Ongoing Practice

1. Use a ruler to measure and mark each shape to the nearest tenth of a centimeter. Then calculate the area.

a.
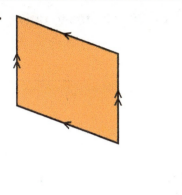

b.
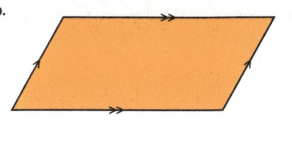

2. Complete two inequalities to match each pan balance.

a.
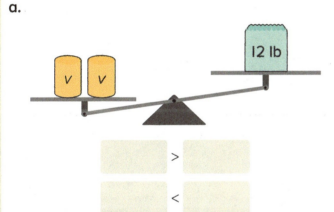

☐ > ☐

☐ < ☐

b.
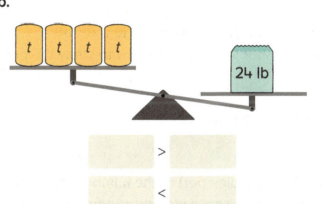

☐ > ☐

☐ < ☐

Preparing for Module 10

Write each division expression as an improper fraction, then as a mixed number.

a. $6 \div 4$ is equivalent to $\frac{6}{4}$ or $1\frac{1}{2}$

b. $8 \div 3$ is equivalent to ☐ or ☐

c. $12 \div 10$ is equivalent to ☐ or ☐

d. $7 \div 5$ is equivalent to ☐ or ☐

e. $9 \div 2$ is equivalent to ☐ or ☐

f. $26 \div 8$ is equivalent to ☐ or ☐

9.3 Algebra: Identifying the range of possible values for an inequality

Step In

How would you represent each statement on a number line?

| The temperature on Monday was less than 30°F. | The pizza took less than 30 minutes to arrive. |

Alisa drew this number line.
Which story does it match? How can you tell?

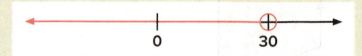

> The part of the line that shows the range of possible values goes beyond 0. There's only one story that can allow negative values.

What number line would you draw to represent the other statement?
What would be different about this number line?

There are fewer than 12 miles to the city.
What number line can you draw to show the possible values?

Archie drew this number line.

Why does the pink part of the number line stop where it does?
Why is the circle solid instead of empty at the position of zero?

Step Up

1. Read each statement. Then circle the possible values.

a. A box of granola cereal has a mass of less than 30 oz.	b. Evan counts backward by 2 from 29.
12 32 20 –16	5 –13 18 0

c. A hot air balloon is more than 2,000 ft above sea level.	d. The day's high temperature is higher than the low temperature of –7°F.
350 5,340 –400 2,070	–10 0 –5 –15

326 ORIGO Stepping Stones · Grade 6 · 9.3

2. Read each statement. Using two different colors, draw a number line to show the range of possible values. Then list three possible values.

   a. There is less than $4\frac{1}{2}$ oz of cheese left in the refrigerator.

   b. The shipwreck is more than 200 meters below sea level.

   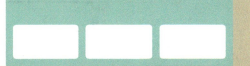

3. Identify whether the number line indicates a limit to the **maximum** possible values. Then circle the values that are possible.

   a.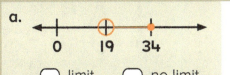

   ○ limit   ○ no limit

   25   34
   49
   19   13

   b. number line: -5 (open circle), 0 (closed dot), 8

   ○ limit   ○ no limit

   -3   -5
   2
   8   -7

   c. number line: 0, 9 (open circle), 86

   ○ limit   ○ no limit

   86   51
   9
   236   47

   d. number line: 0.2, 0.7 (open circle), closed dot, 2

   ○ limit   ○ no limit

   1.96   0.7
   -0.8
   2   1

4. Identify whether the number line shows a limit to the **minimum** possible values. Then circle the values that are possible.

   a. number line: -7, 0, 21 (open circle)

   ○ limit   ○ no limit

   -7   10
   21
   -25   2.9

   b. number line: 3/8 (closed dot), 2, 3 (open circle)

   ○ limit   ○ no limit

   $2\frac{1}{8}$ $\frac{3}{8}$
   2
   0   3

   c. number line: -41, 0 (closed dot), 50 (open circle)

   ○ limit   ○ no limit

   32   0
   -37
   50   0.5

   d. number line: -164, 0 (open circle), 81

   ○ limit   ○ no limit

   -1   0.03
   0
   -200   65

Step Ahead   Choose one inequality from Question 3 or 4. Then write a statement to match.

9.4 Algebra: Working with inequalities

Step In What situation could you write to match this inequality? $g < -4$

Which of these number lines represents the inequality? How do you know?

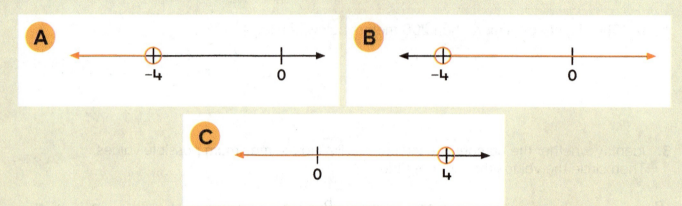

Write an inequality to match each of the other number lines.

Number line ☐   Number line ☐

In January, the daily temperature in Sacramento, California can range from 39°F to 54°F.

In the same month, the daily temperature in Ozark, Arkansas can range from 25°F to 48°F.

Which temperatures could both places share on a day in January?

Step Up

1. Label each number line to show which inequality it represents. Cross out the inequality that is not represented.

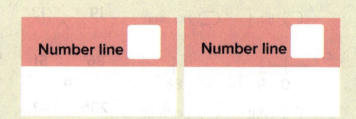

A $h < -6$ B $g < 6$ C $6 < p$ D $k > -6$

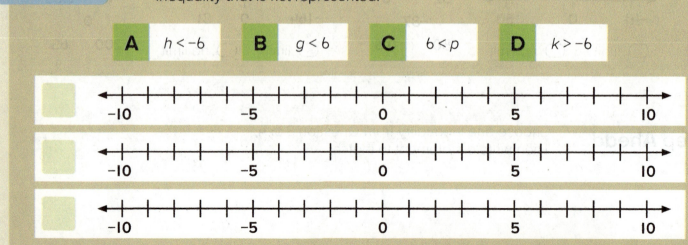

2. Shade three inequalities. Draw a number line to represent each inequality you shaded. Then write a word problem to match.

$r < 3.6$ $d < 45$ $-11 > w$ $h > 100,000$ $n < -4$

a.

b.

c.

Step Ahead Draw a number line to show the possible solutions.

Jayden and Bella are driving to the same university in separate cars. Jayden has fewer than twelve miles to go. Bella has fewer than five miles to go. They both have the same distance left to travel. How many miles could they each have left to travel?

9.4 Maintaining concepts and skills

Think and Solve   Each space is numbered in this diagram.

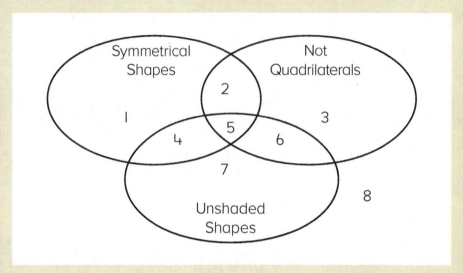

Write a number to show where each shape belongs.

a.   b. △   c. ▭   d. ●   e. ⌂

Words at Work Write a word problem that could match this number line. Note whether the number line shows a limit to the possible values. Write two possible values to match the number line.

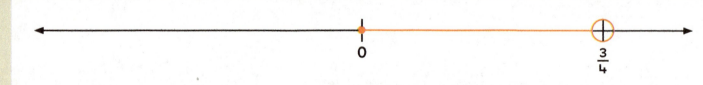

Ongoing Practice

1. Calculate the area of each shape below. Remember to record the correct units.

a.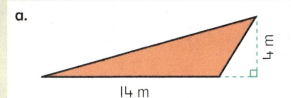

b.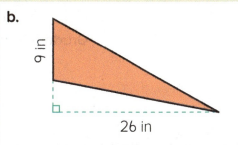

2. Draw a number line to show each inequality.

a. $r < 15$

b. $38 > f$

c. $x > 10$

Preparing for Module 10

Write the absolute value to complete each equation. Use the number line to help.

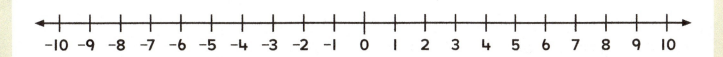

a. $|-4| =$ ☐

b. $|5| =$ ☐

c. $|-10| =$ ☐

d. $|0| =$ ☐

e. $|-8| =$ ☐

f. $|3| =$ ☐

9.5 Statistics: Introducing statistics

Step In What do you notice about this dot plot?

A **dot plot** is also known as a **line plot**.

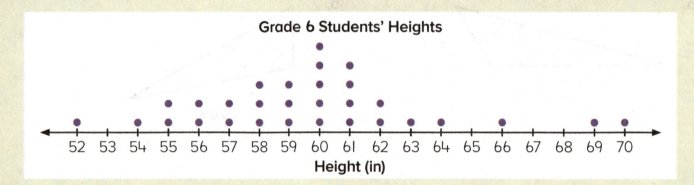

What data is represented on this dot plot?
What question could this graph answer?

It could help answer the question, What are the heights of students in Grade 6?

A **statistical question** is one that can have a variety of answers.
The data above came from a statistical question.
What question could you ask the class with answers that will vary?

If I ask everyone their name, the answers will vary, but I don't think the data would be easy to record except as a list.

Step Up 1. Explain why each of the following questions is or is not a statistical question.

a. How long does it take Kuma to walk to school?

b. How do students in Grade 6 get to school?

c. What distances do Grade 6 students travel to school?

d. How far away does Callum live from school?

2. Use the dot plot on page 332 to answer the following questions.

   a. How tall is the shortest person in Grade 6?

   b. What is the difference between the heights of the tallest and shortest students?

   c. What is the most common height?

   d. What fraction of the students were shorter than 60 in?

   e. Henry was away on the day the data on students' heights were collected. He is 67 inches tall. How would you describe Henry's height compared to those of the other students?

3. a. What is a statistical question you can ask the class now?

   b. With your teacher's help, collect and record data about your own or someone else's question.

   c. What do you think would be a good format to show the data? Why?

Step Ahead For what other reasons could the data displayed on page 332 be useful? Explain your answer on page 356.

9.6 Statistics: Identifying the mode

Step In

Anya was comparing the reading level of a Grade 2 book against a Grade 6 book. She counted the number of letters in the first 26 words of each book and graphed the results.

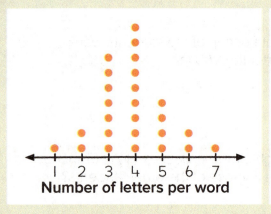
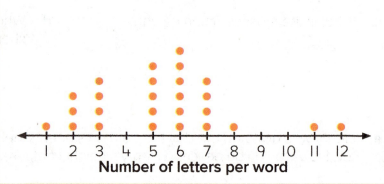

What do you notice about each of the two graphs?

Which graph do you think shows the Grade 6 data? Why?

> The value that occurs most often in a data set is called the **mode**.

Step Up

1. Look at the dot plot on the right above and answer the following questions.

   a. What is the mode?

   b. What fraction of the words are above the mode?

   c. What fraction of the words are below the mode?

   d. Based on your answers for the previous two questions, do you think the mode is a good measure for that data set? Explain your answer.

2. Imagine you own a market stand that sells rings. You would like to make sure you have variety but also not have too many sizes that won't sell well.

   Select a finger on one hand to measure. Write your hand and finger selection below.

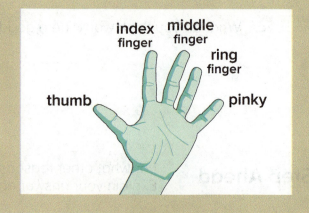

3. Use millimeters to measure the widest part of the finger you chose.

4. Work with your teacher to collect the measurement for that same finger for each student. Record the measurements below. Remember to include your own.

5. Create a dot plot below to show the data you collected in Question 4.

6. Write your answers to the following on page 356.

   a. Explain which ring size you would stock the most.
   b. Explain which ring size you would stock the least.
   c. If you wanted to stock 100 rings at all times, how many rings of each size would you have?

Step Ahead

A Grade 6 class had an event for parents and students. Sketch a possible dot plot that shows the ages of all the people attending. It does not have to be exact. Use page 356 if more space is needed.

9.6 Maintaining concepts and skills

Computation Practice

★ To reveal a fun fact, use a ruler to draw a line to the correct quotient. The line will pass through a letter. Write each letter above its matching quotient at the bottom of the page. Some letters are used more than once.

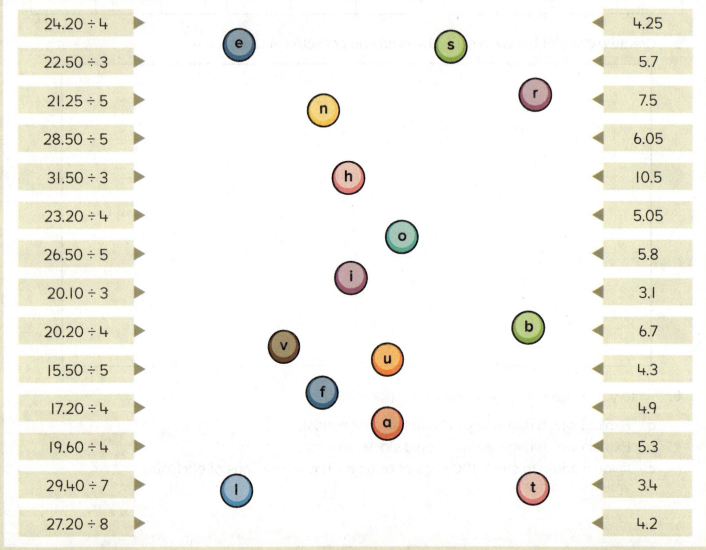

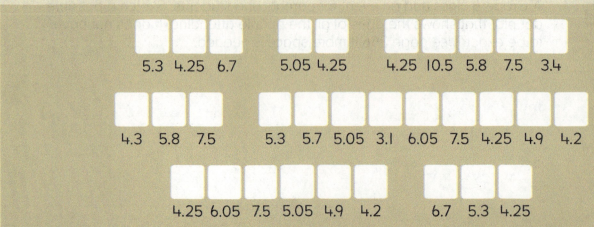

Ongoing Practice

1. Draw extra lines if needed to adjust each shape below. Then calculate the area. Remember to include the appropriate units.

a.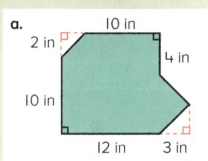

b.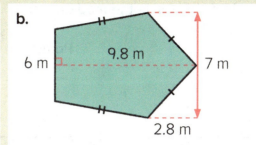

2. Read each statement. Then circle the values that are possible.

a. A water cooler has a volume that is less than 150 fl oz.	b. Hernando's top score on his computer game was greater than 1,578.
−18 80 162 23	1,099 −500 2,300 1,790
c. The day's maximum temperature is more than the minimum temperature of −5°F.	d. Anna counts backward by 5 from 43.
−12 −4 −8 7	28 30 −2 19

Preparing for Module 10

The table shows the race times of some students in seconds. Complete the dot plot to show the results.

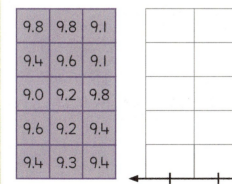
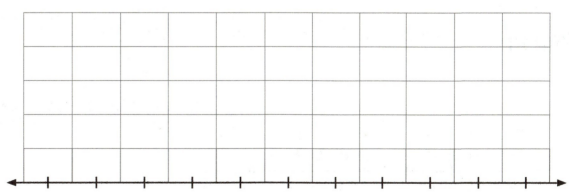

9.7 Statistics: Calculating the median

Step In

Marvin records his basketball team's first five results of the season on a dot plot. The mode cannot be identified at this stage, so he considers another measure called the **median**. The median is the middle value when the data is arranged in order. Circle the median score.

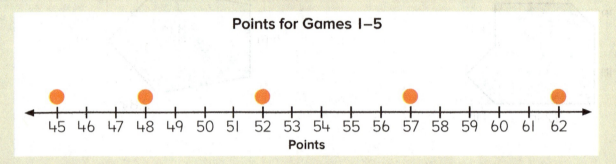

Marvin then records his team's remaining results of the season on the same dot plot.

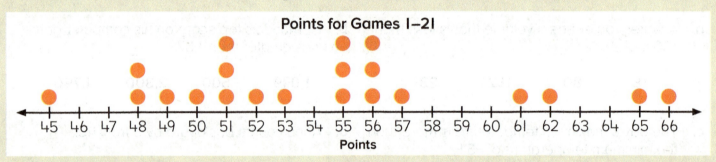

Look at the data for all 21 games. What do you notice? Can you identify the mode?

Circle the new median. How is it different from before? How is it the same?

How useful is looking at the mode and median after 5 games?

How does this compare to looking at the mode and median after 21 games?

I can see more than one mode. The median has also shifted from where it was after 5 games.

Step Up

1. Determine the median of each data set. Remember that values must be in order.

a.

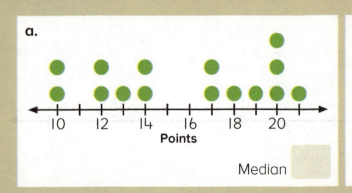

Median ▢

b. 1, 14, 12, 5, 14, 8, 10, 16, 13, 6, 14, 9, 13

Median ▢

2. Your teacher will provide a paper ball for you to conduct an experiment. Record your result to the nearest half a foot using a decimal fraction.

3. Work with your teacher to collect measurements for your class. Record the measurements below. Remember to include your own.

4. Create a dot plot below to show the data you collected in Question 3.

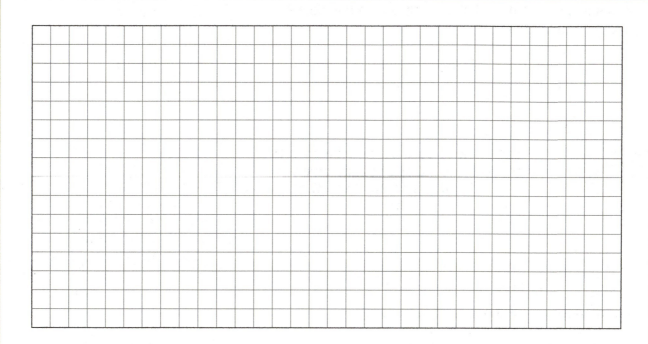

5. Answer the following questions. Use page 356 for Question 5c.

a. What is the mode for the class?

b. What is the median for the class?

c. If you had to arrange targets on the floor, at what distances would you place them, and what scores would each target show? Explain your answer by referring to the mode and median.

Step Ahead A data set has 19 values. The median of the data set is 15. The mode is 9. Write a possible list of values in the data set.

ORIGO Stepping Stones · Grade 6 · 9.7 339

Step Up

1. This table shows how much rain fell over different periods of time in different towns. Calculate the mean daily rainfall for each town.

Town	Total rainfall (to $\frac{1}{4}$ inch)	Days recorded	Mean daily rainfall (inches)
A	3.25	10	
B	7.50	20	
C	1.50	10	
D	9.25	100	
E	9.75	30	

Working Space

2. Which town's mean daily rainfall is the best representation of its typical rainfall? Which is the worst? Explain your thinking.

3. Calculate the mean of each set of numbers.

a. 5   5   10   10   10   10   15   15   20   20   Mean ☐

b. 5   5   10   10   10   10   15   15   20   1,000   Mean ☐

4. In Question 3, which mean is the better representation of its set of numbers? Explain your thinking.

Step Ahead

1. Shade the ○ beside all the contexts where calculating the mean would **not** be a useful measure.

○ The mean of a series of barcode numbers.
○ The mean shoe size of a town's population.
○ The mean of test scores for a single person.
○ The mean of house numbers in a street.
○ The mean of household income in a town.
○ The mean of a group of phone numbers.

2. Explain your choices on page 356.

9.8 Maintaining concepts and skills

Think and Solve

While he was on a plane, Peter counted the number of adults, children, and empty seats.

- $\frac{1}{5}$ of the seats were occupied by children.
- $\frac{3}{4}$ of the seats were occupied by adults.
- 10 seats were empty.

a. How many passenger seats were on the plane?

b. Write how you figured it out.

Words at Work

Write the answer for each clue in the grid. Use words from the list. Some words are not used.

Clues Across

4. A measure that represents the center of a data set.
5. The middle value when the data is arranged in order.
6. Inequalities can be ___ by using common factors.
7. A relationship between two quantities that are not equal.

Clues Down

1. A ___ question is one that can have a variety of answers.
2. The value that has occurred most often in a data set.
3. The ___ is calculated by adding all the numbers in a data set together, then dividing the sum by the number of values.

Word list:
- average
- relationship
- mode
- median
- equality
- statistical
- data
- mean
- variable
- simplified
- inequality
- balanced

Ongoing Practice

1. Complete each equation. Show your thinking.

a. $\frac{2}{3} \div \frac{1}{5} =$ ☐

b. $\frac{8}{10} \div \frac{4}{6} =$ ☐

c. $\frac{5}{8} \div \frac{2}{6} =$ ☐

d. $\frac{2}{3} \div \frac{5}{8} =$ ☐

2. Explain why each of the following questions is or is not a statistical question.

a. How many minutes do sixth grade students spend working on their homework each night?

b. How long did Chloe spend reading on Monday night?

c. How many school days are there in the year?

d. What is the most popular sport played by sixth grade students?

Preparing for Module 10

This dot plot shows the mass of some puppies born at an animal shelter in March. Use the dot plot to answer the questions.

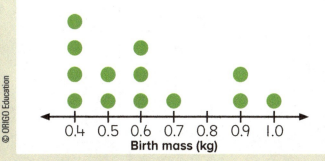

a. How many puppies weighed less than 0.6 kg? ☐

b. How much lighter was the lightest puppy than the heaviest puppy? ☐

c. What is the total mass of the puppies that weighed more than 0.8 kg? ☐

9.9 Area: Using nets to calculate surface area of prisms

Step In

Brady wants to make a toy tent out of paper for his younger brother. He drew a picture to help him determine the amount of paper he would need.

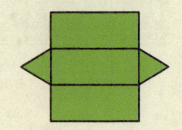

Brady also drew a picture of what the tent would look like when flattened out. This type of picture is called a **net**.

If each rectangular part has the same area, how can you determine the amount of paper Brady will need?

Calculate the area of each segment of the tent. Then add them together to find the **surface area** of the entire object.

> I can see each rectangle is the same size. So are the two triangles. That should help make the calculations faster.

How could you calculate the area of the same tent if Brady did not need to include the floor in his design?

Step Up

1. Beside each picture of a 3D object write the letter of the matching net. Some nets have no match.

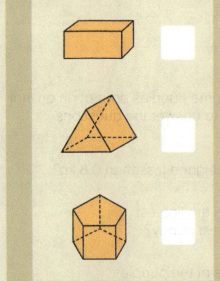

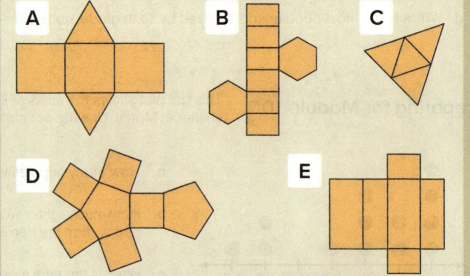

9.8 Statistics: Calculating the mean

Step In Eva recorded the snowfall in her town for 9 days in January.

How can you calculate the median?

Do you think the median is representative of the snowfall so far?

Eva wondered what the data would look like if the same amount of snow had fallen each day but the total amount of snow after 9 days was still the same.

How much snow would have fallen each day in this case? How could you calculate it?

Day	Depth (to $\frac{1}{2}$ in)
Day 1	0.5
Day 2	0.5
Day 3	4.0
Day 4	7.5
Day 5	5.0
Day 6	0.0
Day 7	0.0
Day 8	1.0
Day 9	4.5

First I'll need to know how much total snow there was. That might be easier if I look for pairs of numbers that make a whole number sum.

The **mean**, or **average**, is calculated by adding all the values in the data set together, then dividing the sum by the number of values.

Calculate the mean for the 9 days Eva recorded.

total snow ÷ number of days = mean snowfall

☐ ÷ ☐ = ☐

The table on the right shows the next 11 days of snow.

Build on from the previous total amount of snow to calculate the total for all 20 days. Then calculate the mean.

total snow ÷ number of days = mean snowfall

☐ ÷ ☐ = ☐

Day	Depth (to $\frac{1}{2}$ in)
Day 10	5.0
Day 11	6.5
Day 12	4
Day 13	2
Day 14	1
Day 15	1.5
Day 16	1.5
Day 17	0
Day 18	0
Day 19	0.5
Day 20	2.5

2. Label the lengths on the net of this prism.

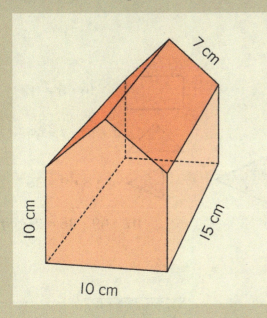
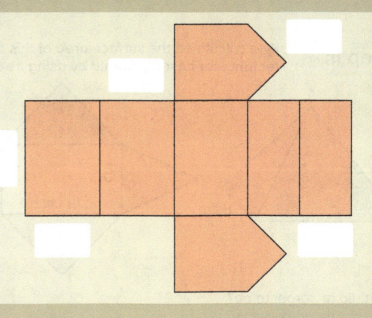

3. Use the same color to show the surfaces that have the same area. Then calculate the surface area of each object. Show your thinking. Remember to include appropriate units.

a. *Note*: All edges of this object are the same length.

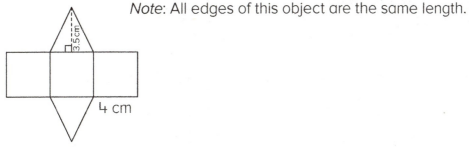

b.

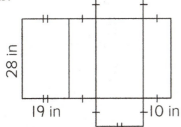

Step Ahead

A teacher has five identical prisms, each with dimensions as shown. The prisms need repainting and she has some tiny jars of paint. If each jar covers 200 cm², how many jars will she need to paint all five objects? Show your thinking on page 356.

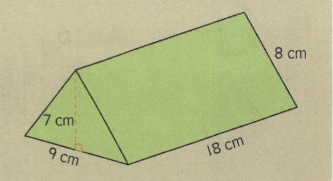

9.10 Area: Using nets to calculate surface area of pyramids

Step In

Kyle calculated the surface area of this rectangular-based pyramid by using its net.

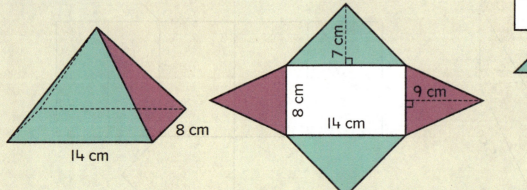

$14 \cdot 8 = 112$

$\frac{1}{2} \cdot 14 \cdot 7 = 49$

$\frac{1}{2} \cdot 8 \cdot 9 = 36$

$112 \cdot 49 \cdot 36 = 197 \text{ cm}^2$

What did he forget to do?
What did he do correctly?

How could you calculate the surface area of this pyramid made from equilateral triangles?

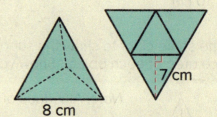

Step Up

1. Beside each picture of a 3D object write the letter of the matching net. Some nets have no match.

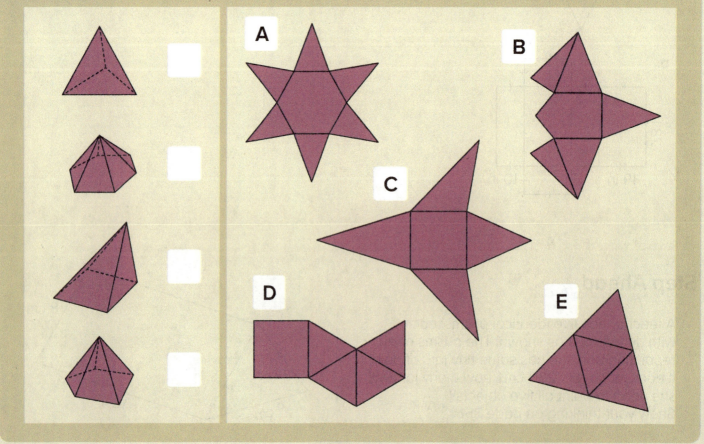

2. Label the lengths on the net of this pyramid.

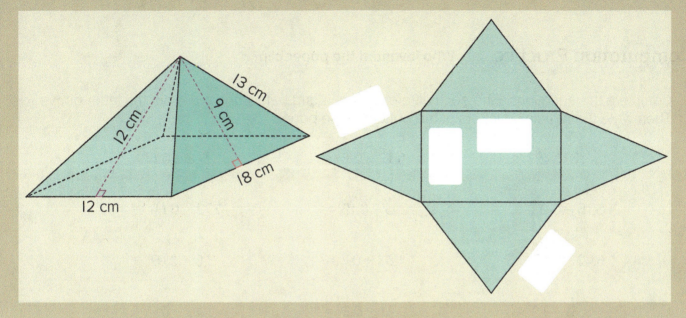

3. Use the same color to show the surfaces that have the same area. Then calculate the surface area of each object. Show your thinking. Remember to include appropriate units.

a. 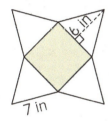 Note: All edges of this object are the same length.

b.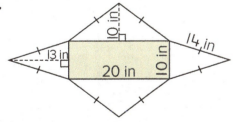

Step Ahead

Carol designed a net for a pyramid that has one hole in each face. Each large triangle is equilateral with a side length of 6 in and perpendicular height of 5 in. Each hole is an equilateral triangle with dimensions that are half those of the larger triangle. What is the surface area of the pyramid? Show your thinking.

9.10 Maintaining concepts and skills

Computation Practice — Who invented the paper clip?

★ Complete the equations. Find each difference in the grid below and cross out the letter above. Then write the remaining letters at the bottom of the page.

11.9 − 2.3 = 9.6	4.12 − 0.6 = 3.52	4.8 − 0.6 = 4.2
5.15 − 0.03 = 5.12	3 − 0.15 = 2.85	0.72 − 0.64 = 0.08
6.42 − 4.02 = 2.4	1.63 − 0.7 = 0.93	3.5 − 0.07 = 3.43
3.81 − 1.9 = 1.91	2.1 − 0.3 = 1.8	2.24 − 0.4 = 1.84
4.7 − 2.01 = 2.69	5.36 − 1.9 = 3.46	5.4 − 0.09 = 5.31
1.7 − 1.32 = 0.38	2.4 − 0.5 = 1.9	0.88 − 0.4 = 0.48
3.5 − 0.7 = 2.8	1.25 − 0.3 = 0.95	2.1 − 0.03 = 2.07
1.27 − 0.9 = 0.37	0.84 − 0.2 = 0.64	5.18 − 0.6 = 4.58
5.69 − 2.3 = 3.39	5.2 − 0.03 = 5.17	1.9 − 0.5 = 1.4
6.4 − 5.17 = 1.23	1.38 − 0.5 = 0.88	4.75 − 0.2 = 4.55

H	S	W	O	S	M	T	A	R	N
5.12	4.05	9.6	0.88	1.8	3.52	4.2	1.45	2.8	2.85
M	H	I	P	U	Q	A	I	L	L
5.23	2.69	1.91	0.38	0.99	0.93	1.84	4.58	0.37	2.4
B	W	A	E	B	L	E	R	B	D
0.95	3.43	1.23	0.82	5.31	2.86	3.46	1.4	3.33	0.64
J	F	Y	A	G	L	Y	O	V	R
0.08	2.7	3.39	4.5	2.07	0.48	1.95	1.9	5.17	4.55

Write the letters in order from the ✱ to the bottom-right corner.

| S | A | M | U | E | L | | B | . | | F | A | Y |

Ongoing Practice

1. Use the invert-and-multiply method to write equivalent expressions. You do not have to write the quotient.

a. $\frac{4}{5} \div \frac{2}{3}$

b. $\frac{5}{8} \div \frac{2}{6}$

c. $\frac{9}{12} \div \frac{2}{5}$

d. $\frac{4}{9} \div \frac{4}{6}$

2. Use the dot plot to answer the questions.

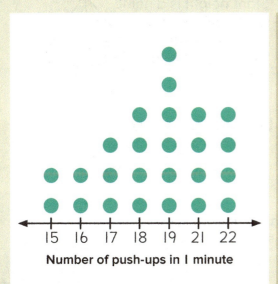

Number of push-ups in 1 minute

a. What is the mode?

b. What fraction of the results are above the mode?

c. What fraction of results are below the mode?

d. Do you think the mode is a good measure for this data set? Explain your answer.

Preparing for Module 10

Each small cube in these prisms is 1 cm³. Calculate the volume of each prism. Show your thinking.

a.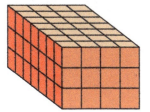

b.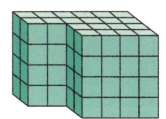

9.11 Area: Calculating surface area of prisms and pyramids

Step In

The Louvre Museum in France has a large glass pyramid in its courtyard. The base is a square with a side length of 34 m and each triangular face has a slant height of 27 m. There is no glass on the base of the pyramid. What area does the glass cover?

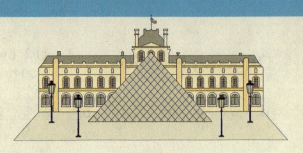

Label the pictures below to show the measurements described above.

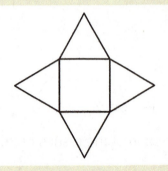

Slant height is a name for the perpendicular height of a triangle when it is sloping.

What steps would you follow to find the total amount of glass on the pyramid at the Louvre?

What expression could you write?

I know a square-based pyramid has four congruent faces. This makes calculating the surface area easier.

Step Up

1. Write a single expression to show how you would calculate the surface area of each object. You do not need to simplify the expression.

a.
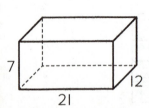

b.
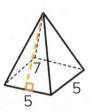

c.
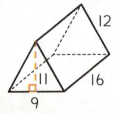

d.
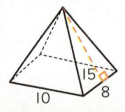

◆ 350

2. Calculate the surface area of each of the objects in Question 1. All measurements are in feet. Show your thinking.

a.

b.

c.

d.

3. Jose is calculating the total surface area of an object. He writes this expression.

$$(3 \cdot 5) + (3 \cdot 5) + (3 \cdot 6) + (3 \cdot 6) + (5 \cdot 6) + (5 \cdot 6)$$

a. For what type of object is he calculating the surface area? Explain your thinking.

b. Write a simpler expression to calculate the surface area of the same object.

Step Ahead

Imagine this square-based pyramid is cut in half from the apex to the base at a perpendicular angle. What is the surface area of one half of the pyramid?

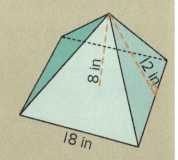

9.12 Area: Solving word problems

Step In

Lomasi wants to cover a box in decorative paper. She wants to draw a net of the box to help calculate the total amount of paper she will need.

The length of the box is 24 cm and its height is half its width. Its width is a third of its length.

Draw and label a net for the box.

What is the most efficient way for her to calculate the surface area of the box?

What expression could she write to show this?

Step Up

1. Solve each problem. Show your thinking.

   a. Terek is repainting the inside of his rectangular pool. The pool is 18 ft long, 10 ft wide, and 8 ft deep. The water level is $\frac{1}{2}$ ft below the edge of the pool. What is the internal surface area of the pool?

   b. A square-based pyramid is made as a special gift box. The base is 6 in wide and the slant height of each triangular face is $\frac{10}{12}$ the length of the base. It can hold 30 items inside. What is the surface area of the box?

2. Read the questions about this cottage and calculate the answers. Show your thinking.

a. The roof of a cottage needs refurbishing. One side of the roof is 30 ft long by 12.5 ft wide. If a bundle of shingles can be purchased to cover 24 ft², how many bundles are needed?

b. The vertical surfaces of the cottage need to be painted. The front and back of the cottage are the same shape. The cottage is 21 ft wide and the length is 2 ft less than the length of the roof. The side walls are 7 ft high, and the height from the bottom of the door to the peak of the roof is 14 ft. Ignoring the doors and windows, what is the area to be painted?

Step Ahead

Two identical triangular-based pyramids have their bases stuck together completely to make what is called a dipyramid. The triangular faces are equilateral with a side length of 18 cm, and the slant height of each triangle is 12 cm. What is the surface area of the outside of the dipyramid?

9.12 Maintaining concepts and skills

Think and Solve

Use the clues to write which rectangle has the greatest area.

Clues
- The length of Rectangle Y is 6 cm. This is twice the length of Rectangle X.
- The length of Rectangle Z is 3 cm. This is twice the width of Rectangle Y.
- The width of Rectangle X is 2 cm. This is $\frac{1}{3}$ the width of Rectangle Z.

Working Space

Words at Work

Britney is building this hen house. The part on the grass has no chicken wire on the bottom. The wire she needs to buy is 1.5 yd wide and is sold in whole or half yard lengths.

Explain in words how to calculate the amount of wire she will need to buy so there is the smallest amount possible left over.

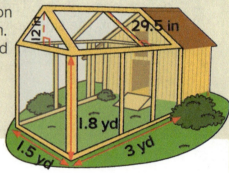

Ongoing Practice

1. Use the invert-and-multiply method to solve each division problem.

a. $\frac{3}{7} \div \frac{2}{5} =$

b. $\frac{8}{9} \div \frac{3}{5} =$

c. $\frac{3}{5} \div \frac{2}{8} =$

d. $\frac{4}{5} \div \frac{2}{12} =$

e. $\frac{4}{12} \div \frac{2}{5} =$

f. $\frac{4}{9} \div \frac{2}{3} =$

g. $\frac{7}{8} \div \frac{1}{2} =$

h. $\frac{4}{5} \div \frac{5}{6} =$

i. $\frac{2}{6} \div \frac{1}{2} =$

2. Determine the median of each data set. Remember that values must be ordered.

a. 12, 10, 8, 12, 4, 12, 9, 16, 15

b. **Spelling Test Results**

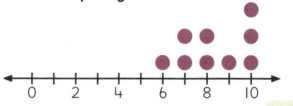

Preparing for Module 10

Solve each problem. Show your thinking and include the appropriate unit in your answer.

a. Jack and Emma are sending gifts to relatives overseas. They share the cost of shipping and the space within the box equally. The box is 24 in by 36 in by 18 in. How much space is available for each person?

b. Abigail is packing the same size boxes into a storage unit. The unit is 6 ft by 12 ft by 10 ft. The boxes measure 4 ft by 3 ft by 3 ft. Abigail has put 5 boxes into the unit already. How much space is left in the unit?

Working Space

10.1 Statistics: Measuring variability using mean and absolute deviation

Step In

Two groups of students recorded the number of hours they each spent playing online games over one week.

Set A: 1, 1, 3, 3   Set B: 0, 0, 7, 1

What do you notice about the mean for each set of data?

Is the mean a good way to describe how the two data sets are different?

> To **deviate** is to depart from a particular point or path.

In statistics, it is sometimes useful to consider the spread of the data. One way to do this is by calculating the **mean absolute deviation** (MAD).

The mean for the data in Set A is 2.
How is the mean shown in this display?

How does the display show the distances between the data points and the mean?

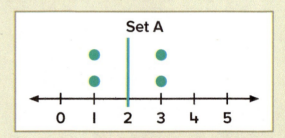

The distance from the mean to each data point can be called the **deviation**. It is an absolute value.

Isaac uses this table to record each deviation.

How does he calculate the deviation of each value from the mean?

Isaac now calculates the mean deviation by dividing the sum of the deviations (4) by the number of values (4).

What is the mean deviation?

Value	Distance from mean
1	1
1	1
3	1
3	1
Total	4

Step Up

1. a. The mean for the data in Set B is 2. Use this number line to calculate the MAD. Show your thinking.

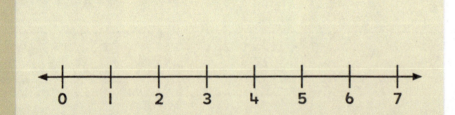

MAD is ____

b. Compare the mean absolute deviation for data sets A and B. What does the difference indicate about the data?

358

2. Calculate the mean for each data set. Then use the number line to help calculate the MAD. Show your thinking.

a. 5, 6, 7, 8, 9, 10, 11, 12, 13

Mean is

MAD is

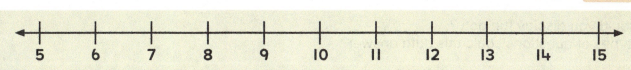

b. 10, 10, 12, 6, 8, 8

Mean is

MAD is

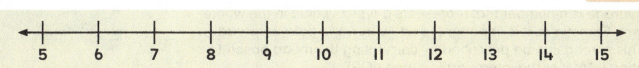

3. The examples in Question 2 show the test results of two groups of students from Grade 6. Each group is given the same test. Which group of students performed more consistently on their test? Think about the deviation around each mean to justify your decision.

Step Ahead

Circle the dot plot that represents the data set with the greatest variation around the mean. Do not carry out any calculations. Then explain how you formed your decision.

A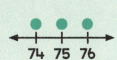

B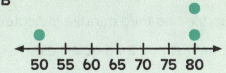

C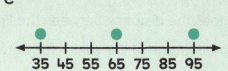

10.2 Statistics: Measuring variability using quartiles and interquartile range

Step In

Some students are asked about the number of hours that they spend on social media each week. The results are then recorded.

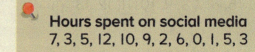

Hours spent on social media
7, 3, 5, 12, 10, 9, 2, 6, 0, 1, 5, 3

How could you display the data?
What types of questions could this data answer?

Knowing the average number of hours spent on social media might be helpful.

The **average** is a number that represents a typical value in the whole data set. Sometimes it is important to know how the data is spread out. This spread can be described by calculating the **mean absolute deviation (MAD)** or the **interquartile range (IQR)**.

Cole follows these steps to calculate the IQR of the social media data.

First he orders the data to find the median.

0, 1, 2, 3, 4, 5, | 5, 6, 7, 9, 10, 12

Then he finds the median of each half. These medians are called **quartiles**.

0, 1, 2, | 3, 4, 5, | 5, 6, 7, | 9, 10, 12
    first   second   third
  quartile   quartile quartile
   (2.5)   (5)   (8)

The **interquartile range (IQR)** is calculated by finding the difference between the first and third quartiles.

IQR   third quartile − first quartile
IQR   8 − 2.5 = 5.5

What does the distance between the first quartile and third quartile indicate?

What percentage of the values fall between the first and third quartiles?

Step Up

1. Order this data from least to greatest. Then write the first, second, and third quartiles.

3, 6, 2, 1, 7, 12, 9

1st quartile ☐   2nd quartile ☐   3rd quartile ☐

◆ 360   ORIGO Stepping Stones · Grade 6 · 10.2

2. Order the data from least to greatest. Write the first, second, and third quartiles. Then calculate the IQR.

a. 3, 15, 20, 21, 8, 7, 16, 5, 11, 9

1st quartile ☐   2nd quartile ☐   3rd quartile ☐   IQR ☐

b. 15, 20, 10, 10, 5, 25, 30, 20

1st quartile ☐   2nd quartile ☐   3rd quartile ☐   IQR ☐

3. Calculate the quartiles and IQR for the data on this dot plot. Show your thinking.

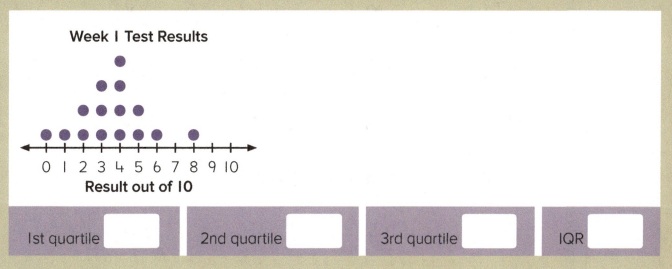

1st quartile ☐   2nd quartile ☐   3rd quartile ☐   IQR ☐

Step Ahead

A test with 10 questions was given to 20 students. This data was collected from the final scores.

1st quartile is 5.
2nd quartile is 7.
3rd quartile is 8.
IQR is 3.

Complete the dot plot to match all the data.

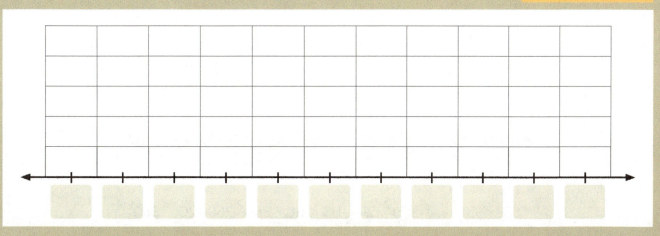

10.2 Maintaining concepts and skills

Computation Practice

★ Complete the equations. Then write each letter above its matching answer at the bottom of the page to discover a science fact. Some letters are used more than once.

$\frac{1}{4}$ of 180 =		e
$\frac{1}{5}$ of 275 =		b
$\frac{1}{2}$ of 720 =		o
$\frac{1}{5}$ of 250 =		t
$\frac{1}{10}$ of 360 =		g
$\frac{1}{4}$ of 460 =		n
$\frac{1}{4}$ of 820 =		i
$\frac{1}{4}$ of 260 =		y
$\frac{1}{2}$ of 520 =		r

$\frac{1}{4}$ of 360 =		h
$\frac{1}{10}$ of 750 =		m
$\frac{1}{4}$ of 412 =		v
$\frac{1}{5}$ of 620 =		a
$\frac{1}{2}$ of 450 =		w
$\frac{1}{10}$ of 330 =		l
$\frac{1}{5}$ of 350 =		d
$\frac{1}{3}$ of 600 =		s

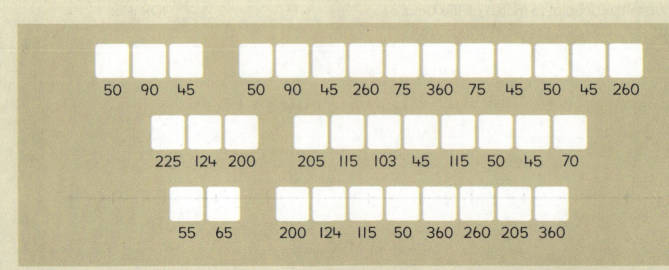

50 90 45   50 90 45 260 75 360 75 45 50 45 260

225 124 200   205 115 103 45 115 50 45 70

55 65   200 124 115 50 360 260 205 360

Ongoing Practice

1. Calculate each percentage. Show your thinking. Remember to include the appropriate unit in your answer.

a. 30% of $2.50 = ☐

b. 50% of 4.5 mi = ☐

c. 20% of 3 lb = ☐

d. 80% of $1.50 = ☐

2. Calculate the mean for each data set. Then use the number line to help calculate the mean absolute deviation (MAD). Show your thinking.

a. 5, 5, 6, 6, 7, 8, 9, 10

Mean is ☐

MAD is ☐

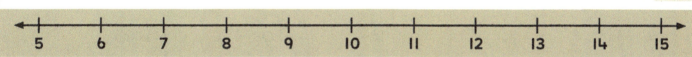

b. 15, 14, 14, 12, 10, 10, 8, 5

Mean is ☐

MAD is ☐

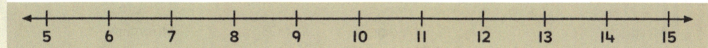

Preparing for Module 11

Read the ratio for each table. Then write numbers to complete each table.

a. The ratio of pears to apples is 2:5.

Pears	2				
Apples	5				

b. The ratio of white paint to blue paint is 3:1.

White paint	3				
Blue paint	1				

10.3 Statistics: Introducing box plots

Step In

Look at this list of data values.

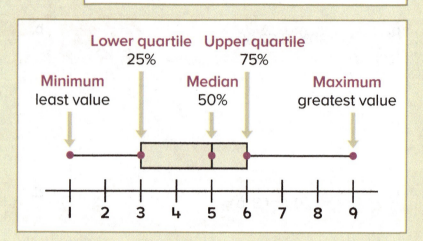

Lisa draws this box plot to show the data.

Describe how the two representations are similar.

What are the lower quartile (Q1) and upper quartile (Q3)? How did she calculate these values?

What does the difference between Q1 and Q3 indicate about the data?

What are the least and greatest values? What do these values indicate?

What does a box plot not show?

Step Up

1. Interpret each box plot. Then write the five-number summary.

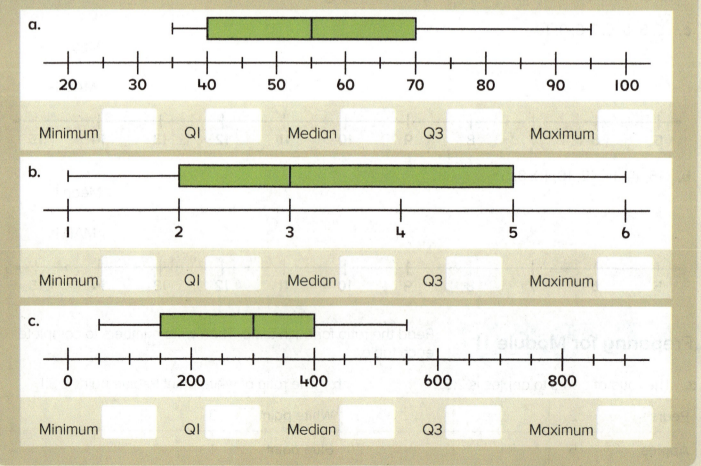

2. Match each description to a box plot in Question 1. One of the descriptions has no match.

a. Number of televisions in one household

b. Results from a classroom assessment

c. The number of calories in Grade 5's morning snacks

d. The ages of students in Grade 8

3. a. A class of Grade 2 students collected data on the number of pets they own. Create a box plot to show the data. Then complete the five-number summary.

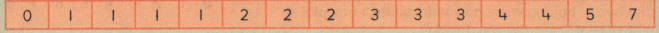

Minimum ☐ Q1 ☐ Median ☐ Q3 ☐ Maximum ☐

b. How do you calculate the percentage of the class that own two pets?

4. Competitors are challenged to see how many apples they can hold without dropping any. This box plot shows the results. Do the results match your expectations? Explain.

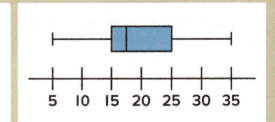

Step Ahead Write a data set to match this box plot.

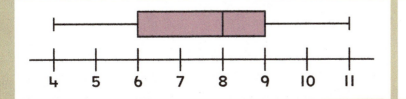

10.4 Statistics: Consolidating box plots

Step In

Two groups of homeowners were surveyed about the number of hours that they spend gardening each week.

The first group of homeowners lived in the city. The second group lived in the country.

These box plots show the results.

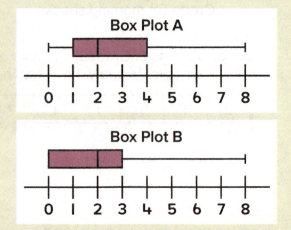

Which group of homeowners do you think each box plot might represent? How did you decide?

Compare the two box plots. What are the similarities and differences?

How would you describe the variation around each median?

Step Up

1. The following high temperatures (°F) were measured in a city in the United States from January 1 to January 15.

| 73 | 74 | 73 | 74 | 81 | 84 | 84 | 79 | 71 | 72 | 74 | 76 | 70 | 74 | 79 |

a. Order the data from least to greatest. Then complete the five-number summary.

Minimum ☐   Q1 ☐   Median ☐   Q3 ☐   Maximum ☐

b. Draw a box plot to show the data.

c. Write a short report to describe the variation in temperature. Make sure you describe the location of the city and the state where the data might have been collected.

2. Fifteen students in each of the three Grade 6 classes were asked how long they spent playing outdoors each week.

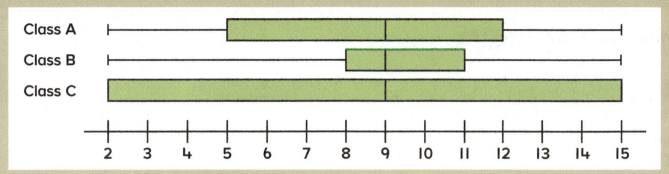

a. Describe the variation between any two of the classes.

b. Write **A**, **B**, or **C** to show which dot plot matches each class result. Use the shape of the data to help you decide.

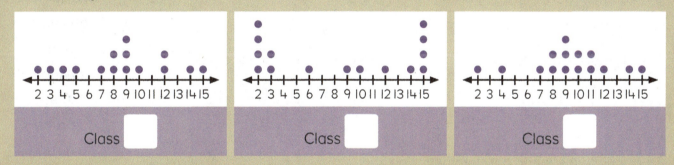

Class ☐   Class ☐   Class ☐

c. For what purpose could the dot plot be more useful than the box plot?

d. For what purpose could the box plot be more useful than the dot plot?

Step Ahead When looking at the results of the class surveys in Question 2, Luis concluded that *Class C clearly spent more time outdoors than students in the other two classes*. Do you agree or disagree with Luis? Explain your thinking to another student.

10.4 Maintaining concepts and skills

Think and Solve

What is **10%** of **20%** of **30%** of **40%** of **1,000**? ▢

Write how you calculated the answer.

Words at Work

Use words from the list to complete true sentences.

a. Calculating the mean absolute ▢ is a way to analyze the variation between each data point and the ▢.

b. The distance from the ▢ to each data point is called the deviation. It is an ▢ value.

c. To calculate the mean absolute deviation, the ▢ of all deviations is ▢ by the total number of values in the data set.

d. The ▢ is a number that represents a typical value in the whole data set. In some cases, it is important to know how the data is spread out. This ▢ can be described by calculating the mean ▢ deviation, or the ▢ range.

e. The interquartile range is calculated by finding the ▢ between the ▢ and ▢ quartiles.

Word list:
deviation
mean
interquartile
absolute
sum
divided
first
spread
average
difference
third

Ongoing Practice

1. Calculate each percentage. Show your thinking.

a. 5% of 200 =

b. 9% of 150 =

c. 18% of 250 =

d. 22% of 300 =

e. 15% of 900 =

f. 2% of 15 =

2. Order the data from least to greatest. Write the first, second, and third quartiles. Then calculate the interquartile range (IQR).

a. 7, 3, 10, 10, 8, 13, 10, 4, 8, 9

1st quartile   2nd quartile   3rd quartile   IQR

b. 35, 20, 15, 5, 15, 40, 35, 30, 25

1st quartile   2nd quartile   3rd quartile   IQR

Preparing for Module 11

Solve each problem. Show your thinking.

a. Wendell walked 1,200 yards in 10 minutes. In 15 minutes Naomi walked 1,875 yards. Who walked faster?

b. Ruben makes juice using blueberries and strawberries in a ratio of 3:2. Donna makes juice using grapes and strawberries in a ratio of 5:3. Whose juice will have a stronger strawberry taste?

10.5 Statistics: Introducing histograms

Step In

A new basketball court is being built at school. The principal wants to find a fair height to hang the new hoop and decides to investigate the heights of students who like to play basketball. The results are shown in the frequency table and histogram below.

Students in Grades 1 to 6 Who Like Basketball

Height (in)	Frequency	Total
40–44	||||	4
45–49	𝍤 𝍤 ||	12
50–54	𝍤 𝍤 𝍤 𝍤 𝍤 𝍤	30
55–59	𝍤 𝍤 𝍤 𝍤 𝍤 𝍤 𝍤 ||||	39
60–64	𝍤 𝍤 𝍤 𝍤 𝍤 𝍤 ||	32
65–69	𝍤 𝍤 ||||	14
70–74	𝍤 |||	8

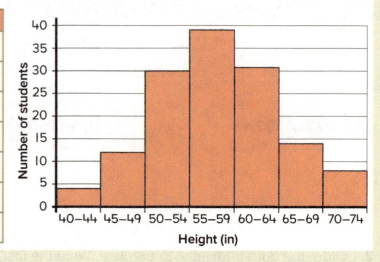

What do the results tell you about the student heights?

What is similar about the **frequency table** on the left and the **histogram** on the right?

The student heights have been grouped. These groups are called **class intervals**.

The tallest bar has the most frequencies. It's called the **modal class**.

How could the principal use this data to inform her decision about how high to place the basketball hoop?

Step Up

1. Customers at a gaming store are asked to write down their age. The manager wants to show the results in a frequency table. Write the class intervals that the manager could use. Then draw the tally marks.

Age	Tallies

17	14	17	14	23	16	15
15	27	16	27	28	22	25
20	13	22	20	18	17	19

2. The histogram shows the number of available car spaces in a city parking lot. Complete the table to match the histogram.

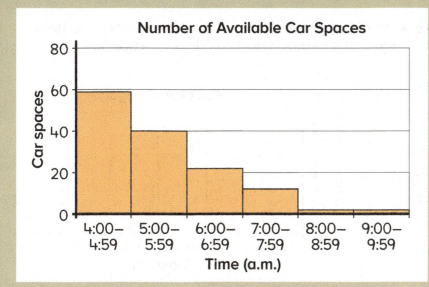

Time (a.m.)	Frequency
4:00–4:59	

3. Interpret the above displays to answer each question.

a. What is the difference between the number of spaces available between 5:00 a.m. and 5:59 a.m., and 7:00 a.m. and 7:59 a.m.?

b. How many more spaces are available between the hours of 4:00 a.m. and 4:59 a.m. than 9:00 a.m. and 9:59 a.m.?

c. The parking attendant wants to know the average number of car spaces that are available during this morning period. Explain the steps that he could follow.

Step Ahead

Draw bars on the histogram to show what you think the number of available car spaces would be across the afternoon and evening. Then write a short justification for the shape of the graph.

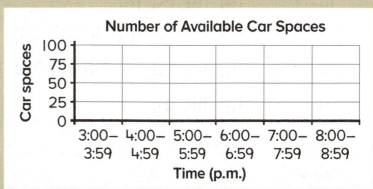

10.6 Statistics: Analyzing and creating histograms

Step In An airline records the weight of baggage being checked for a flight. One item is checked by each passenger. The results are shown in this histogram.

What can you share about the mass of the baggage?

The checked baggage limit is 50 lb. How many passengers exceeded this limit?

One passenger's bag weighed exactly 39.8 lb. How is the mass of their bag represented in this histogram?

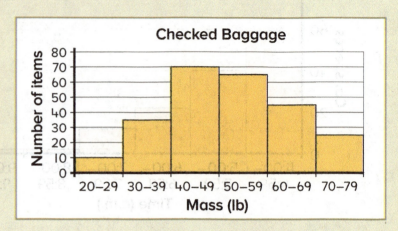

The second bar from the left includes all baggage that weighs exactly 30 lb and everything less than 40 lb. So a bag weighing 39.8 or even 39.998 would be included in this bar.

Which measure of center (mean, median, or mode) would you use to describe the weight of the checked baggage?

How could you calculate the mean or median weight? What additional data would you need?

You need to know the exact weight of each passenger's bag to calculate the mean or median. This histogram shows the **modal class**. The data is called **bi-modal** if two bars are the same or similar in height.

Step Up 1. Complete the table.

Checked Baggage		
Weight (lb)	Frequency	Total
10–19	卌 卌 卌 卌 l	21
20–29	卌 卌 卌 卌 卌 卌 卌	
30–39	卌 卌 卌 卌 卌 卌 卌 卌 ll	
40–49	卌 卌 卌	
50–59	卌 lll	
60–69	ll	

372

2. Use the data from Question 1 on page 372 to complete the graph.

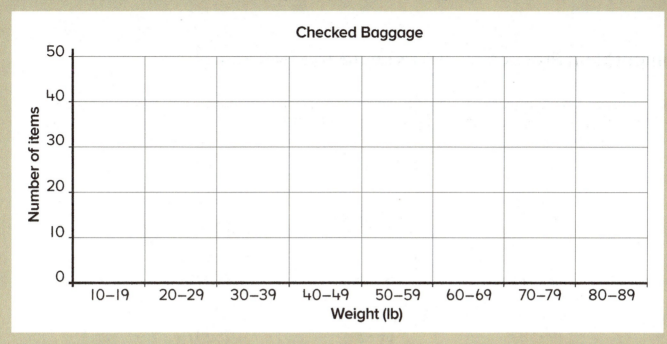

3. Use the table from Question 1 and the histogram above to complete each answer.

a. If one bag is checked for each passenger, how many passengers boarded the plane?

b. The baggage limit for this particular flight is less than 40 lb. How many passengers exceeded this limit?

c. What is the modal class?

d. What does the modal class tell you?

Step Ahead The flight captain warns that there is not enough fuel to carry more than $2\frac{1}{2}$ tons of baggage. Using the information in the Step In, should the flight staff be concerned? Remember, 1 ton = 2,000 lb.

10.6 Maintaining concepts and skills

Computation Practice

What is a group of parrots known as?

★ Complete each equation. Find each product in the grid below and cross out the letter above. Then write the remaining letters at the bottom of the page.

25 • 26 =	15 • 62 =	12 • 55 =
15 • 12 =	32 • 25 =	22 • 15 =
42 • 15 =	16 • 25 =	35 • 22 =
15 • 18 =	65 • 6 =	25 • 17 =
18 • 25 =	35 • 19 =	54 • 10 =
23 • 15 =	15 • 24 =	25 • 22 =
48 • 15 =	16 • 15 =	35 • 28 =
35 • 14 =	25 • 12 =	
46 • 20 =	50 • 26 =	

H	A	P	K	S	M	T	E	A
630	180	295	240	360	1,300	400	450	760
N	T	I	D	E	Q	U	A	I
985	300	660	110	210	800	390	665	330
B	M	A	O	B	L	E	R	N
490	435	920	730	650	270	770	345	470
J	A	I	O	G	U	M	O	V
980	550	605	930	720	870	310	425	540

Write the letters in order from the ✱ to the bottom-right corner.

Ongoing Practice

1. Complete each calculation. Show your thinking on the double number line.

a. I spent $42 which is 30% of what I earned.

I earned _____ .

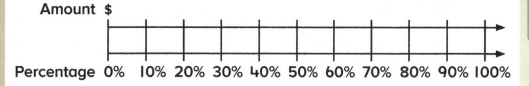

b. I spent $18 which is 60% of what I earned.

I earned _____ .

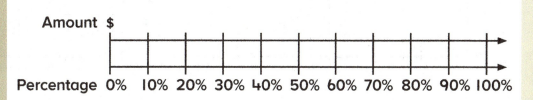

2. Interpret each box plot. Then write the five-number summary.

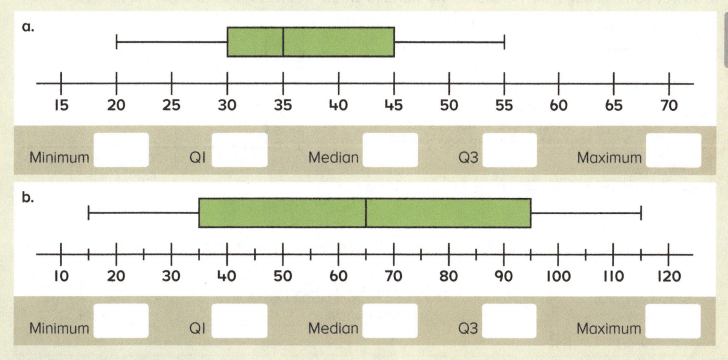

a. Minimum ____ Q1 ____ Median ____ Q3 ____ Maximum ____

b. Minimum ____ Q1 ____ Median ____ Q3 ____ Maximum ____

Preparing for Module 11

Solve each problem. Show your thinking and include the appropriate unit in your answer.

a. Carrina spends 80 minutes doing homework each night. 35% of that time is spent reading. How long does Carrina spend reading?

b. Chayton had $320 in savings. He went shopping and spent 70% of his savings. How much did Chayton spend?

10.7 Statistics: Working with histograms

Step In Jamar compares these two histograms. He says that Histogram A has less variation because it looks flatter.

Do you agree with him?

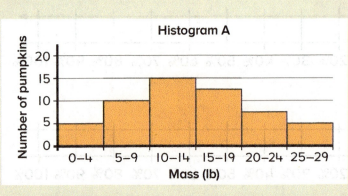

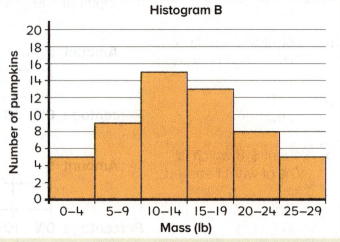

Isabelle suspects that her clothing store is more popular among certain age groups. She decides to survey some shoppers in her store. She starts to create this histogram to show the results.

What is something unusual about what she has made?

The age range in each group is not consistent. This means that more shoppers fall within certain age groups.

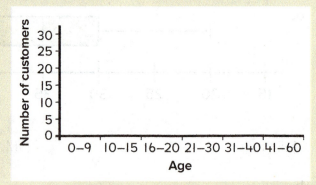

What other intervals could she use to group the ages?

Step Up 1. This table shows the age of some pay-per-view subscribers. Complete the table.

Pay-per-view Subscriptions		
Ages	Frequency	Total
10–19	𝍬	5
20–29	𝍬 𝍬 𝍬 |||	
30–39	𝍬 𝍬 𝍬 𝍬 𝍬 𝍬 𝍬 𝍬 ||	
40–49	𝍬 𝍬 𝍬 𝍬 𝍬 𝍬 𝍬 𝍬 𝍬 ||||	
50–59	𝍬 𝍬 𝍬 𝍬 𝍬 ||||	
60–69	𝍬 ||	
70–79	|||	
80–89		

376

2. Use the data from Question 1 on page 376 to complete the graph.

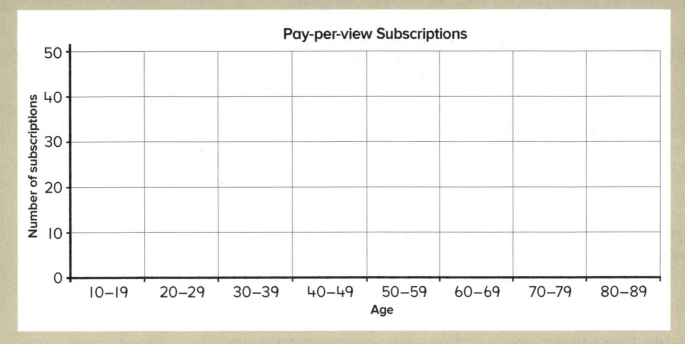

3. What does the data tell you about the age of customers who have pay-per-view subscriptions?

4. Imagine you work for a company that live streams sporting events. How could the results influence the marketing and advertising decisions that you make?

Step Ahead

Subscribers of a new social media website were asked to share their age. A summary of the results is shown below. Complete the histogram to match the results.

50 participants were surveyed.

50% of the participants were <20 years of age.

Participants aged 15-19 formed the modal class.

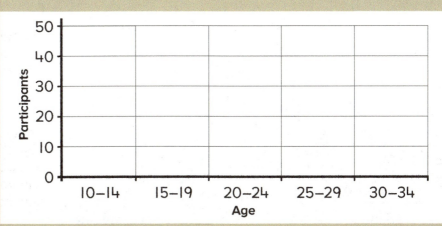

10.8 Volume: Reviewing volume

Step In

Connor wants to find out which one has the greater volume. How could he calculate and compare the two volumes?

He decides to pack centimeter cubes along the bottom of each box. After he finishes the first layer, what could he do next to make the packing process faster?

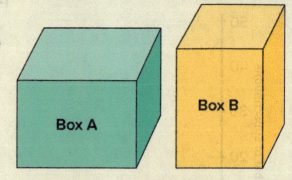

Box A

Box B

Ruth reminds him that he could use a formula for calculating volume. She suggests these two different formulas.

$$V = l \cdot w \cdot h$$
$$V = b \cdot h$$

What do the variables represent in each formula?

Why do these formulas work to find the volume of rectangular prisms?

How does the second formula relate to the cube packing that Connor started?

Connor measured each dimension and used the formula to find the volume of each box.

What has he not included in his answer?
How would you explain this mistake to him?

Box A	Box B
$V = 6 \cdot 3 \cdot 10$	$V = 9 \cdot 7 \cdot 3$
$= 18 \cdot 10$	$= 63 \cdot 3$
$= 180$ cm	$= 189$ cm

Connor did not use the right units. Centimeters are only a measure of length. He needs to record the number of cubic centimeters that would fit inside each box.

How would you calculate the volume of this prism?

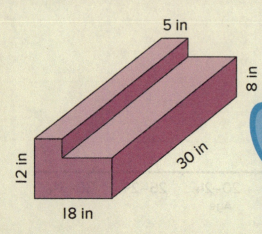

5 in
8 in
30 in
12 in
18 in

Some dimensions are missing but I should have enough information to figure out what they are.

I could split the prism into two objects and calculate the volume of each. I could also calculate the volume of a larger object and subtract the excess amount.

Step Up

1. Use a formula to calculate the volume of each prism. Show your thinking.

a.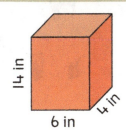

b.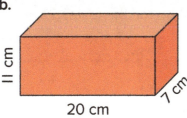

c.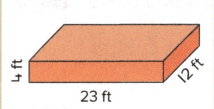

2. Solve the problem. Show your thinking on page 394. | 1 cm = 10 mm |

   Selena finds some pieces of plywood with dimensions marked on them.
   Two are 80 mm by 190 mm, two are 19 cm by 34 cm, and two are 34 cm by 8 cm.

   a. What is the volume of the box that can be made with all the pieces?

   b. What is the surface area of the finished box?

3. Calculate the volume of this building. Show your thinking.

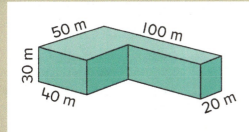

Step Ahead   Look at the objects on the right.

a. Which object has the greater volume?

b. Which has the greater surface area?

c. Explain your thinking on page 394.

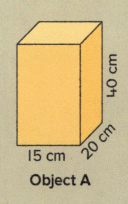
Object A

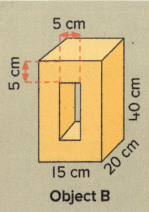
Object B

10.8 Maintaining concepts and skills

Think and Solve Who ran the fastest?

- Antonio ran 2,000 m in 9 minutes.
- Rita ran 2,500 m in 12 minutes.
- Kasem ran 1 km in 7 minutes.
- Vishaya ran 1.2 km in 6 minutes.

Working Space

Words at Work

A group of people were surveyed to determine the time period in which they watched the most television. Look at the information in the histogram and write about what you notice. How do you think the histogram would change if it included a range after midnight to before 5 a.m.?

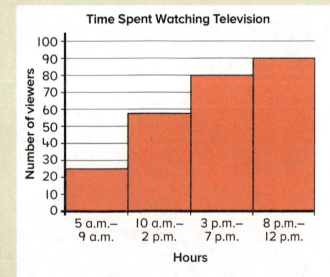

Ongoing Practice

1. This table shows how much snow fell over different periods of time in different towns. Calculate the mean daily snowfall for each town.

Town	Total depth (to $\frac{1}{4}$ inch)	Days recorded	Mean depth (inches)
A	2.50	10	
B	3.25	20	
C	7.50	30	
D	11.75	100	
E	6.25	10	

Working Space

2. This histogram shows the height of trees in an orchard. Complete the table to match the histogram.

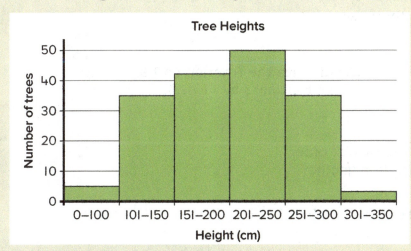

Height (cm)	Frequency
0–100	

Preparing for Module 11

Measure and label the dimensions of each shape in centimeters. Then calculate the area of each shape. Remember to include the correct unit.

a.
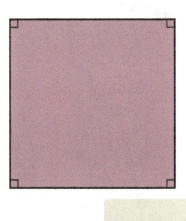

b.
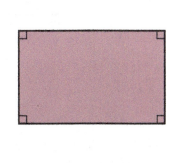

c.
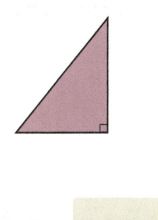

10.9 Volume: Rectangular-based prisms with one fractional side length

Step In

A raised platform for a band measures 8 yards by 3 yards, and is half a yard high. The space underneath the platform is used for storage. How could you calculate the volume of the storage space?

Susan imagined the platform in whole cubic yards then split it in half to show the fraction. She then used this idea to calculate the volume.

$3 \cdot 8 \cdot 1 = 24$
$24 \cdot \frac{1}{2} = 12$ yd³

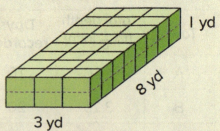

How could you calculate the volume of the platform if it was $2\frac{1}{2}$ yd high?

The formulas $V = l \cdot w \cdot h$ and $V = b \cdot h$ can be shortened to $V = lwh$ and $V = bh$.

I would first calculate 3 by 8. That gives 24. Then I'd multiply 24 by 2, then 24 by $\frac{1}{2}$, then add those two products together.

I would write the fraction as 2.5. Then I'd multiply 3 by 8 by 2.5.

What other methods could you use? Is there a particular order of multiplication that you think would be easier than another?

How could you find the volume of a container measuring 5 inches by 6 inches by $1\frac{1}{4}$ inches?

Step Up

1. Calculate the volume of each object. Show your thinking.

a.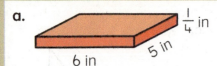
 6 in, 5 in, $\frac{1}{4}$ in

b.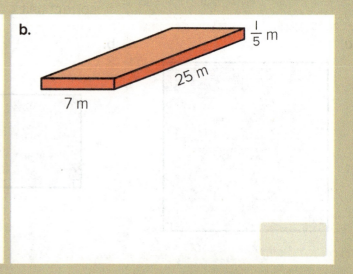
 7 m, 25 m, $\frac{1}{5}$ m

382

2. Calculate the volume for each situation. Use common fractions and mixed numbers in your calculations and answers.

   a. A wooden block that is 8 in long, $\frac{3}{4}$ in high, and has a width of 2 inches.

   b. A piece of lumber that is 6 ft long, 1 foot wide, and $\frac{1}{12}$ ft thick.

3. Calculate the volume for each situation. Use decimal fractions in your calculations and answers.

   a. Mulch that is needed to cover an area of 150 m² to a depth of 0.2 m.

   b. The dirt that is dug from a hole measuring 3 ft deep, 2 ft wide, and 4.1 ft long.

Step Ahead A block of cheese is shaped like a rectangular-based prism. It is cut in half along a diagonal line. What is the volume of one wedge of cheese as shown below?

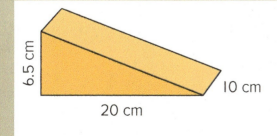

10.10 Volume: Rectangular-based prisms with two fractional side lengths

Step In

Look at these cubes.

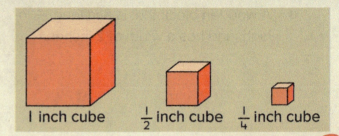

1 inch cube $\frac{1}{2}$ inch cube $\frac{1}{4}$ inch cube

How many $\frac{1}{4}$ inch cubes can fit inside the 1-inch cube? How do you know?

A container is 5 inches by $\frac{1}{2}$ inch by $\frac{1}{4}$ inch.

How many $\frac{1}{4}$ inch cubes will fit inside the container? How do you know?

I could figure out how many $\frac{1}{4}$ inch cubes fit in a 1 inch section then multiply that by 5.

How could you use the information about the number of $\frac{1}{4}$ in cubes to calculate the volume of the container?

I know that there are sixty-four $\frac{1}{4}$ inch cubes in one cubic inch. That means the volume of the container is $\frac{40}{64}$ in³, which is equivalent to $\frac{5}{8}$ in³.

Ben used a formula for calculating volume and wrote $V = 5 \cdot \frac{1}{2} \cdot \frac{1}{4} = \frac{5}{8}$ in³.

If the box was 5 in by $2\frac{1}{5}$ in by $\frac{3}{8}$ in, how could you calculate the volume? What types of fractions would you use? Explain your thinking.

I can change $2\frac{1}{5}$ to a decimal fraction easily but I'm not sure about $\frac{3}{8}$.

$2\frac{1}{5}$ would be easier to work with if I changed it to a common fraction.

Step Up

1. Calculate how many $\frac{1}{4}$ inch cubes will fit in each box. Then write the volume.

a.

$\frac{3}{4}$ in, 2 in, $\frac{1}{4}$ in

_____ cubes   _____ Volume

b.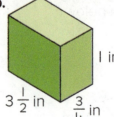

$3\frac{1}{2}$ in, $\frac{3}{4}$ in, 1 in

_____ cubes   _____ Volume

2. Calculate the volume for each situation. Use common fractions and mixed numbers in your calculations and answers.

a. A mattress that is 3 ft wide and $6\frac{1}{4}$ ft long. It is $\frac{2}{3}$ ft thick.

b. A small apartment block with a floor area of $278\frac{1}{5}$ yd² and height of $10\frac{1}{2}$ yd.

3. Calculate the volume for each situation. Use decimal fractions in your calculations and answers.

a. A CD case that is 14.2 cm wide, 1 cm thick, and 12.5 cm high.

b. A cell phone that has dimensions of 14.6 cm by 7 cm by 0.9 cm.

Step Ahead The volume of a rectangular-based prism is $2\frac{2}{6}$ ft³. Write a set of possible dimensions for the prism so that two dimensions are proper fractions. If you need extra space show your thinking on page 394.

length: width: height:

10.10 Maintaining concepts and skills

Computation Practice
Where is the smallest bone in the human body?

★ Complete each equation. Find each answer in the grid below and cross out the letter above. Then write the remaining letters at the bottom of the page.

20% • $6.50 = $1.30	50% • $13.48 = $6.74	25% • $29.96 = $7.49
10% • $19.80 = $1.98	20% • $17.50 = $3.50	50% • $27.22 = $13.61
25% • $3.48 = $0.87	10% • $21.40 = $2.14	20% • $26.55 = $5.31
50% • $17.72 = $8.86	25% • $1.04 = $0.26	10% • $3.70 = $0.37
25% • $18.96 = $4.74	10% • $58.30 = $5.83	50% • $1.98 = $0.99
20% • $37.75 = $7.55	25% • $37.12 = $9.28	10% • $27.70 = $2.77
50% • $9.88 = $4.94	20% • $8.65 = $1.73	25% • $19.80 = $4.95
10% • $43.60 = $4.36	50% • $11.02 = $5.51	20% • $2.85 = $0.57

F	I	N	G	E	R	O	R
$3.50	$5.24	$0.77	$8.86	$4.74	$4.36	$0.87	$9.28
F	**O**	**O**	**T**	**H**	**A**	**N**	**D**
$0.99	$4.94	$1.30	$3.55	$3.79	$5.83	$4.95	$5.51
K	**N**	**E**	**E**	**N**	**O**	**S**	**E**
$7.49	$5.31	$9.99	$2.77	$13.61	$0.26	$6.74	$0.38
E	**L**	**B**	**O**	**W**	**A**	**R**	**M**
$0.37	$1.73	$1.98	$2.14	$7.55	$2.59	$5.85	$0.57

Write the letters in order from the ✱ to the bottom-right corner.

Ongoing Practice

1. Write **<** or **>** to make each inequality true.

a. (11 + 22) ÷ 3 ◯ 15 ÷ 2

b. 14 • 3 ÷ 2 ◯ 4 • 6 + 13

c. 5(3.1 + 4.2) ◯ 4 • 10

d. 10(18 − 13) ◯ 7 • 5

e. 3.14 ÷ 0.2 ◯ 3.14 ÷ 0.02

f. 10^2 • 0.12 ◯ 6.14 • 2

g. 4(72 ÷ 9) ◯ 12 • 2.5

h. 18 • 0.2 ◯ 18 ÷ 2

2. Calculate the volume of this building. Show your thinking.

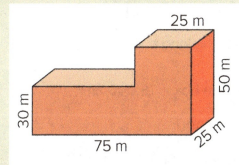

Preparing for Module 11

Complete the table. The base of each object is shaded.

Pyramids			
Number of faces			
Number of vertices			
Shape of base			
Number of sides on base			

10.11 Volume: Rectangular-based prisms with three fractional side lengths

Step In

Identical books are packed into boxes that are 20 cm wide, 30 cm long, and 25 cm high. Exactly 25 books lie flat in each box to fill it completely. What is the volume of one book in cubic centimeters?

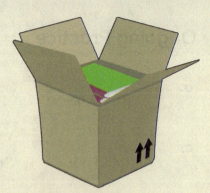

I could calculate the volume of the box then calculate the volume of one book.

I could also find the volume of a single book by dividing one of the box dimensions then multiplying by the remaining dimensions.

Boxes of the books are stacked in layers of 12 and there are 3 layers in total. How many cubic meters is the entire stack? What steps will you follow to calculate the volume?

Jerome showed his thinking this way.

> 100 cm is equivalent to 1 m so the volume of one box is 0.2 • 0.3 • 0.25 = 0.0150 m³, or 0.015 m³.
>
> 3 layers of 12 is 36. V = 36 • 0.015 = 0.54 m³

Morgan showed her thinking this way.

> The volume of one box is 15,000 cm³.
> 3 layers of 12 is 36.
> 36 • 15,000 = 540,000
>
> 100 cm is equivalent to 1 m so the volume of the stack is 540,000 ÷ 100 = 5,400 m³.

What steps did each person use to solve the problem? Whose method is correct?

Step Up

1. Calculate the volume of each prism. Show your thinking.

a.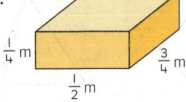

$\frac{1}{4}$ m, $\frac{1}{2}$ m, $\frac{3}{4}$ m

b.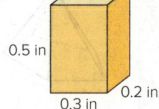

0.5 in, 0.3 in, 0.2 in

2. Calculate the volume for each situation. Use the type of fraction you prefer to calculate and record your answer. Use page 394 if you need extra space to show your thinking.

   a. A brick is 21.5 cm long, 10.25 cm wide, and 6.5 cm high.

   b. A tissue box is $4\frac{1}{2}$ in wide, $3\frac{1}{2}$ in high, and $8\frac{1}{2}$ inches long.

3. Calculate the volume for each situation but take note of the unit that is required in the answer. Use page 394 if you need extra space to show your thinking.

   a. An eraser is 22 mm wide, 45 mm long, and 9 mm thick.

   _____ cm³

   b. A rectangular prism has a base measuring 1.5 ft by 1.25 ft and a height of 0.5 ft.

   _____ in³

Step Ahead

A rectangular prism has a length to width ratio of 4:1. The width and height have a ratio of 3:1. The length of the prism is 24 inches. What are the dimensions and volume of the prism? Show your thinking on page 394.

l = 24 in w = _____ h = _____ V = _____

10.12 Volume: Solving word problems

Step In

Dorothy has a brownie recipe. The mixture fills a 7.5 in by 12 in by 1.5 in pan to the top, but it always spills over the edges when it bakes. So Dorothy has to allow 1 inch between the top of the mix and the pan to allow it to rise.

Would a pan measuring 8 in by 8 in by 3.5 in be more suitable? How do you know? What method would you use to calculate the volume?

If I add an extra inch to the height of Dorothy's original pan I will know what the ideal volume is. Then I can compare the volume of other pans to that amount.

If she had another pan that had a base of 8 in by 10 in, how high would it have to be so that the mixture does not spill over?

Step Up

1. Solve each problem. Show your thinking.

a. A fish tank shaped as a rectangular-based prism is 40 cm long by $12\frac{1}{2}$ cm wide, and 20 cm high. What is the volume of water in it if the tank is $\frac{3}{4}$ full?

b. Robert makes a bread mix that fills a total volume of 360 in³. He has a rectangular pan that is 12 in long and $6\frac{1}{4}$ in wide. What does its height need to be to hold the mixture and allow for just over an inch in rising space? Show your thinking.

2. Solve each problem. Show your thinking.

a. A rectangular fish tank has a base that measures 30 cm by 12.5 cm and is 10 cm tall. It contains a large rock. The tank is filled up to 8 cm. When the rock is removed, the water falls down to a height of 7.5 cm. What is the volume of the rock?

b. A block of wood is 2 cm longer than it is wide. It is 3.5 cm thick, which is 1.5 cm less than its width. What is its volume in millimeters?

Step Ahead   A rectangular-based prism has a volume of 27.28 ft³. Its base has an area of 6.82 ft².

a. What is its height?   **b.** What is its surface area?

10.12 Maintaining concepts and skills

Think and Solve

This large cube is made of smaller cubes. What is the least number of these smaller cubes you would need to make a cube that completely surrounds this one?

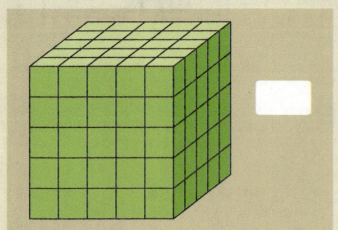

Working Space

Words at Work

Write in words how you would solve this problem.

Amber is setting up her new rectangular fish tank. The tank measures 45 cm by 15.5 cm on the base and is 12 cm tall. Amber purchased some gravel for the bottom of the tank along with two cube-shaped ornaments that were each 5 cm high. To ensure there is enough swimming space for her fish, Amber needs to calculate the volume of the tank once the gravel and ornaments are in place. The gravel fills the bottom of the tank evenly to a height of 2 cm. The tank is 10 cm full. What is the volume of water in Amber's tank?

Ongoing Practice

1. Read the context. Draw a number line to show the range of possible values. Then list three possible values.

a. The box of crackers has a mass that is less than 25 oz.

b. The day's minimum temperature is less than 2°F.

2. Calculate the volume of each object. Show your thinking.

a.

b.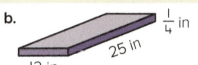

Preparing for Module 11

Complete the table. The base of each object is shaded.

Prisms			
Number of faces			
Number of vertices			
Shape of base			
Number of sides on base			

Working Space

11.1 Ratio: Introducing ratios with three parts

Step In Patricia drew this pattern.

How many times greater is the number of circles than stars?
How many times greater is the number of squares than stars? What other ratios can you describe?

Ratios between three quantities can be written in a similar way to ratios with two quantities.
The ratio between stars, squares, and circles in the pattern above can be represented as 2:8:4.
What is another ratio that could be written to describe the relationship between the three shapes?

Ratios with three parts can be represented in tables of equivalent ratios.

Complete the table below to show how the pattern could be continued.

Number of stars	2				
Number of squares	8				
Number of circles	4				

Step Up

1. For each question choose three different ingredients from the recipe below. Then write each set of three ingredients and the matching ratio.

Ingredients for Carrot Cake	
2 cups sugar	1 teaspoon salt
1½ cups oil	2 teaspoons cinnamon
4 eggs	3 cups carrots
2 cups flour	4 oz walnuts
2 teaspoons baking soda	

a. The ratio between _____, _____, and _____ is ___:___:___.

b. The ratio between _____, _____, and _____ is ___:___:___.

c. The ratio between _____, _____, and _____ is ___:___:___.

2. Use the recipe on page 396 to complete the table below.

Number of eggs	4	8				2
Number of teaspoons of salt	1		6		15	
Number of cups of carrots	3			36		

3. The equipment list below shows some of the items needed to make a single cabinet. For example, one cabinet will use 12 of Item A and 50 of Item B. Use the list to answer the questions. Show your thinking on page 432.

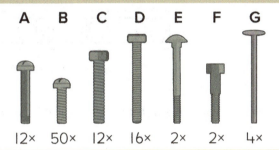

Write the ratios between the following items.

a. Items E, G, and C ☐ : ☐ : ☐

b. Items D, B, and A ☐ : ☐ : ☐

c. Items F, A, and E ☐ : ☐ : ☐

d. Items G, A, and C ☐ : ☐ : ☐

e. There are 24 of Item A.

How many of Item E are needed? ☐

How many of Item C are needed? ☐

f. There are 300 of Item B.

How many of Item G are needed? ☐

How many of Item D are needed? ☐

g. There are 128 of Item D.

How many of Item F are needed? ☐

How many of Item G are needed? ☐

h. There are 50 of Item F.

How many of Item B are needed? ☐

How many of Item A are needed? ☐

Step Ahead A factory packing the cabinet equipment in Question 3 has about 4,200 of each item in stock. How many complete packs of equipment can they make? Show your thinking on page 432.

☐ packs

11.2 Ratio: Using ratios with three parts

Step In

Matthew bought bathroom tiles in three different sizes.

He bought them in the ratio of 2 large, 6 medium, and 14 small.
If the ratio remains constant and he has 528 tiles in total,
how many of each tile does he have?

Paige drew this diagram to help solve the problem.

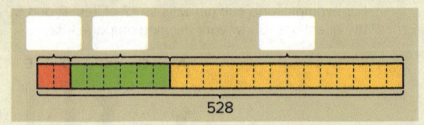

How many parts are there in total?
How can this help you calculate the number of each tile? Complete the missing numbers.

After tiling for a while, Matthew still has 18 large tiles and 54 medium tiles to place.
How many small tiles are there if the ratio remains the same? How many tiles are there in total?

Amy drew this diagram to show the situation.

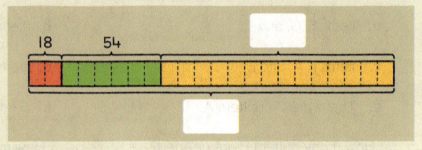

How is the second problem similar to the first one? How is it different? Complete the missing numbers.
What equation would you write to represent finding the unknown amounts in each situation?

Step Up

1. Solve each problem. Show your thinking.

   a. A large bottle of salad dressing has olive oil, vinegar, and honey in the ratio 6:2:1. If a bottle of dressing holds 225 mL of liquid, how much olive oil is there?

   b. Yuma makes a necklace using 6 wooden beads, 3 glass beads, and 6 stone beads. If he has 120 beads in the correct ratio, how many glass beads does he have?

2. Solve each problem. Show your thinking.

a. A recipe for a spice mix includes 3 teaspoons of garlic powder, 2 teaspoons of chili powder, and 4 teaspoons of oregano leaves. If Yasmin makes a large batch of 144 teaspoons of these combined ingredients, how much chili powder will she use?

b. A concrete mixer holds 11 cubic yards of concrete mix that is 2 parts sand, 2 parts gravel, and 1 part cement powder. How much gravel should there be?

c. A tailor sews 7 buttons on each shirt front, 3 on each cuff, and 1 on each side of the collar. If they sew 210 buttons to finish a number of shirts, how many buttons were sewn onto cuffs?

Step Ahead

a. Which equation matches the story in Question 1a on page 398?

○ $(225 \div 3) \cdot 2 = x$ ○ $x = 225 \div 3$ ○ $225 \div [3 \cdot (6 + 2 + 1)] = x$

b. Write a fraction equation that also matches.

11.2 Maintaining concepts and skills

Computation Practice

★ These students were given a math quiz. Check their answers and draw a ✔ beside each correct answer. Add the correct answers for each student and write the score at the bottom of each paper.

Name: Arleen

1. $175 - 30 \div 5 =$ **169** ✔
2. $48 \div 2 \cdot 6 =$ **144** ✔
3. $(3 \cdot 0.3) + 10.5 =$ **19.5**
4. $(3^4) \div 9 + 12 =$ **21** ✔
5. $12 \cdot 2.5 \div 5 =$ **6** ✔
6. $5(8.1 - 7.6) =$ **25**
7. $96 \div 3 \cdot 0 =$ **32**
8. $5^4 - 18 =$ **607** ✔
9. $11(18 - 9) =$ **99** ✔
10. $5 + 250 \div 5 =$ **51**
11. $5^2 - 25 \div (2 + 3) =$ **20** ✔
12. $6 - 15 \div 3 + 1.4 =$ **2.4** ✔
13. $10 \div 2 + (7 - 4) =$ **13**
14. $(9 + 5)^2 =$ **196** ✔
15. $(8 \div 4) \cdot (3 + 7) =$ **20** ✔
16. $(8 + 2)^2 =$ **20**
17. $9 \cdot 8 - 6 - 5^2 =$ **41** ✔
18. $(10 - 3)^2 =$ **1**
19. $2(4 + 3) =$ **14** ✔
20. $8.6 - 2 \cdot 3 - 1 =$ **1.6** ✔

Total correct: ____

Name: Norton

1. $175 - 30 \div 5 =$ **169** ✔
2. $48 \div 2 \cdot 6 =$ **144** ✔
3. $(3 \cdot 0.3) + 10.5 =$ **11.4** ✔
4. $(3^4) \div 9 + 12 =$ **21** ✔
5. $12 \cdot 2.5 \div 5 =$ **6** ✔
6. $5(8.1 - 7.6) =$ **2.5** ✔
7. $96 \div 3 \cdot 0 =$ **0** ✔
8. $5^4 - 18 =$ **2**
9. $11(18 - 9) =$ **99** ✔
10. $5 + 250 \div 5 =$ **55** ✔
11. $5^2 - 25 \div (2 + 3) =$ **20** ✔
12. $6 - 15 \div 3 + 1.4 =$ **4.4**
13. $10 \div 2 + (7 - 4) =$ **8** ✔
14. $(9 + 5)^2 =$ **196** ✔
15. $(8 \div 4) \cdot (3 + 7) =$ **20** ✔
16. $(8 + 2)^2 =$ **100** ✔
17. $9 \cdot 8 - 6 - 5^2 =$ **141**
18. $(10 - 3)^2 =$ **49** ✔
19. $2(4 + 3) =$ **14** ✔
20. $8.6 - 2 \cdot 3 - 1 =$ **18.8**

Total correct: ____

Who had more correct answers? ____

Ongoing Practice

1. Calculate how many $\frac{1}{4}$ inch cubes will fit in each box. Then write the volume.

a.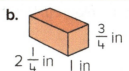

[ ] cubes   Volume [ ]

b.

[ ] cubes   Volume [ ]

2. A bead store is making do-it-yourself bracelet packs. Each pack contains the following items.

A	B	C	D	E	F
14×	15×	9×	6×	5×	2×

Write the ratios between the following items.

a. Items E, D, and C [ ] : [ ] : [ ]   b. Items D, B, and A [ ] : [ ] : [ ]

c. Items F, A, and E [ ] : [ ] : [ ]   d. Items D, A, and C [ ] : [ ] : [ ]

e. There are 56 of Item A.

How many of Item E are needed? [ ]

How many of Item C are needed? [ ]

f. There are 300 of Item B.

How many of Item F are needed? [ ]

How many of Item D are needed? [ ]

Preparing for Module 12

Write and shade equivalent values.

a.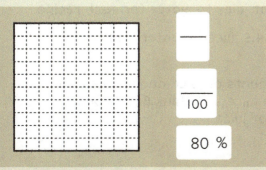

$\frac{}{100}$ 80 %

b.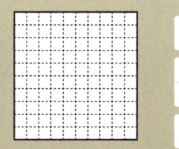

$\frac{2}{100}$ [ ] %

11.3 Ratio: Comparing ratios with three parts

Step In Joel made fruit punch using 3 cups of apple juice, 1 cup of cranberry juice, and 5 cups of sparkling water.

This can be represented using a tape diagram like this.

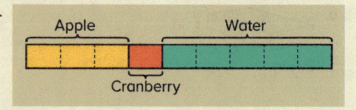

What is the ratio of water to apple?
What is the ratio of water to the total amount of liquid?

How are these two ratios different?

Beth made fruit punch with the same ingredients in different quantities shown below.

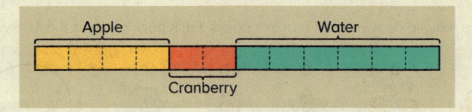

Whose punch will taste more like apple than cranberry? How do you know?

What ratios could you write to prove it?

> I think I need to find an equivalent ratio for one or both of the recipes so I can compare easily.

Whose punch will have a stronger taste of juice overall? How do you know?

What ratios could you write to prove it?

> I need to look at the ratio of juice to water for each recipe.
>
> Joel's recipe has a ratio of 4:5 juice to water and Beth's has a ratio of 6:6.
>
> Beth's recipe has equal amounts of juice and water, and Joel's has more water than juice so I think Beth's punch will have a stronger taste of juice.

Step Up

1. Draw and label a tape diagram to show the ratio for each situation.

   a. Steven makes green paint using 7 cups of white paint, 4 cups of yellow paint, and 8 cups of blue paint.

   b. Sharon makes green paint using 7 cups of blue paint, 6 cups of white paint, and 3 cups of yellow paint.

2. Use the ratios in Question 1 to answer these questions. Show your thinking on page 432.

   a. Whose green paint will look more yellow?

   b. If white paint lightens each mixture, whose green paint will be a lighter shade?

   c. Write a ratio of white to yellow to blue that will make a green paint that is lighter than Steven's paint and darker than Sharon's paint. Use whole numbers only and be sure to name each part of the ratio.

Step Ahead

Refer to Steven's mixture in Question 1. If there are 27 cups of white paint, 14 cups of yellow paint, and 32 cups of blue paint, how many batches or part batches of green paint can Steven make? Show your thinking.

11.4 Ratio: Solving word problems with three-part ratios

Step In

Emilio, Sandra, and Hiro run a small business.

Since they each put different amounts of time and money into the business they agree to share the profits so that Emilio gets 15%, Sandra gets 45%, and Hiro gets 40%.

What ratio could you write to describe the split of profits?

In January the profit is $5,000. What do you need to do to calculate the amount they each get?

In February, Emilio's share of the profit is $1,200. How much do Sandra and Hiro each receive? What do you need to do to calculate the amounts this time?

Step Up

1. Write an equation to match each problem. Use a variable to represent the unknown value. You do not need to solve the problem.

a. The ratio between fat, carbohydrates, and protein in a batch of granola bars is 3:10:1. If there are 39 grams of fat in the batch, how much protein is there?

b. A pair of socks is made of 73% cotton, 25% polyester, and 2% spandex. A pack of 6 socks weighs 30 grams. If the masses of the different fabrics remain the same, what is the total number of grams of polyester and cotton in one sock?

c. A muffin recipe uses 1 cup of diced apple, 3 cups of diced pear, and 2 cups of berries. If a bakery uses 57 cups of pear, how many cups of fruit will have been used in total?

d. Mrs. Jones has 3 tubs of counters. The ratio between the red, green, and yellow counters in each tub is 2:4:1. If there are 96 green counters in one tub, how many red counters are in all tubs?

e. A farmer grows three types of lettuce. For every 5 heads of iceberg lettuce, they grow 3 romaine and 2 red leaf. There are 240 heads of romaine lettuce in each field. How many lettuces in total are in 5 fields?

f. Vincent sells grab bags of 6 comics. In each bag he puts 1 rare comic, 2 semi-rare, and 3 common ones. During one year he sold 63 common comics in grab bags. How many rare comics did he sell?

2. Solve each problem. Show your thinking on page 432.

a. Olivia is cooking multiple batches of berry muffins. There are 12 muffins in a batch and in 4 batches the ratio between cups of sugar, flour, and berries is 3:8:4. If she has 20 cups of berries how many muffins can Olivia make?

b. Michael read that the three most frequent letters in the English language are E, T, and A, with an approximate occurrence of 13%, 9%, and 8% of all letters. Halfway through 800 words Michael counted 52 instances of E. How many instances of A should there be?

c. A social networking site has 5 age ranges of users. The lowest age range is 13–17 which represents 12% of all users. The next range is 18–24 which is 30% of users. The third range is 25–34 which is 22% of users. If there are 20,000 users, how many are 25 years or older?

d. Carmela is knitting a scarf with blue, green, and white yarn. She uses about 200 yards of yarn in total but she uses twice as much blue as white and one-third as much green as blue. About how much green yarn does she use?

Step Ahead

Alexis mixes liquid chemicals A, B, and C in the ratio of 2:6:5. She then mixes them with some water so that chemicals A and B together are 40% of the total amount of liquid. The total amount is 120 mL. How much of chemical B is there? Show your thinking.

11.4 Maintaining concepts and skills

Think and Solve

Oliver has a box of red, blue, and yellow blocks.

- For every 2 red blocks, he has 3 yellow blocks.
- For every 2 blue blocks, he has 3 red blocks.

What is the fewest number of each type of block in the box?

Words at Work

Write in words how you solve this problem.

A farm has chickens, turkeys, cows, goats, horses, and ducks. The ratio of birds to four-legged animals on the farm is 3:1. The ratio of chickens, turkeys, and ducks is 4:2:3. If there are 312 birds and animals in total on the farm, how many of each type of bird are there?

Ongoing Practice

1. Calculate the volume of each prism. Show your thinking.

a.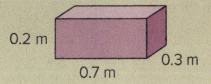

b.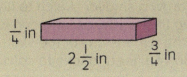

2. Solve each problem. Show your thinking.

a. Felix is planting herbs in his garden. For every 2 coriander seeds he plants 4 oregano seeds and 5 basil seeds. He has planted 121 seeds in total. How many oregano seeds has he planted?

b. Zoe is making candies to sell at the school fair. In each packet she puts in 3 pieces of fudge, 6 marshmallows, and 4 toffees. She has a total of 299 pieces of candy. How many of those are fudge?

Preparing for Module 12

Calculate each percentage. Show your thinking and include the appropriate unit in the answer.

a. 30% of 50 km =

b. 60% of $200 =

11.5 Ratio: Resizing 2D shapes to a given percent

Step In

Anoki is planning a painting for the large rectangular canvas shown below.

He needs to draw a smaller sketch to help with the planning and thinks that a drawing that is 10% as long and wide should work.
What will be the dimensions of the sketch?
What steps do you need to take to calculate them?

2.4 m

3.5 m

Hmmm... I think it might be easier if I change the units to centimeters.

Claire is painting a large canvas like Anoki but it has different dimensions.

She drew a sketch with dimensions of 20 cm and 32 cm to represent it.
If the dimensions are 25% of the full-size canvas, what are the dimensions of the full-size canvas?

What is known and unknown in the first story? How is this different from the second story?

Step Up

1. Measure and label each dimension in millimeters. Then redraw each shape so that each dimension is 40% of the original. Label each new dimension. Show any calculations on page 432.

a.

b.

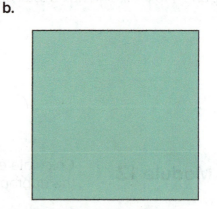

◆ 408

2. Solve each problem. Show your thinking.

a. Amos builds a coffee table that has the same ratio between sides as his rectangular kitchen table. Each dimension is 65% as long as the kitchen table's dimensions. The kitchen table is 50 inches long and 40 inches wide. What are the dimensions of the coffee table?

Length: _____   Width: _____

b. Emilia plans a large garden bed in the shape of a regular hexagon. She wants each side to be 14 ft long but the edging costs too much. So she decides to make each side 80% of that length to reduce the cost. What will be the length of the new perimeter?

_____

c. A copy of a book page is found on a photocopier. The original has been removed but the photocopier shows that it was copied at 70% of the original size. If the photocopied image has a length of 21 cm what was the length of the original book page?

_____

d. The sides of a triangle have all been redrawn at 20% of their original length. If the ratio between the sides is 5:4:8 and the original perimeter of the triangle is 170 inches, what are the dimensions of the new triangle?

Side A: _____
Side B: _____   Side C: _____

Step Ahead

A monument has a ratio of 3:6:4 between its width, height, and length. The dimensions of a model of the monument are 2% of the dimensions of the real monument. If the length of the model is 24 cm, what is the width of the real monument? Show your thinking on page 432 and remember to include the appropriate unit in your answer.

11.6 Ratio: Examining similar rectangles

Step In Tyler has some photos he wants to print.

His phone takes photos that have a length to width ratio of 4:3. He knows from experience that choosing the wrong print size will mean that the photos will not fit the print properly.

Which print size should he choose to get the best results? Why?

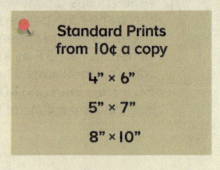

Standard Prints from 10¢ a copy

4" × 6"

5" × 7"

8" × 10"

I don't think any of the print sizes will match exactly, but some are better than others.

What print sizes in whole inches would match Tyler's photos perfectly?

Step Up

1. Cut out all the rectangles on the support page your teacher will give you. Then sort them into groups as directed by your teacher.

2. Paste all the rectangles that belong in Group A below.

3. Paste all the rectangles that belong in Group B below.

4. Describe why all the rectangles in Group A belong together.

5. Describe why all the rectangles in Group B belong together.

Step Ahead

Look at the dimensions below. Determine whether each rectangle belongs to Group A or Group B and write the matching letter beside the dimensions. Write **N** for those with no match.

a. ☐ 10 × 5 cm   b. ☐ 3 × 12 yd

c. ☐ 20 × 12 ft   d. ☐ 7 × 4 in

e. ☐ 16 × 9.6 m   f. ☐ 3.5 × 14 mm

11.6 Maintaining concepts and skills

Computation Practice

Who invented the pedal bicycle?

★ Complete the equations. Find each quotient in the grid below and cross out the letter above. Then write the remaining letters at the bottom of the page.

$4.5 \div 5 =$	$0.6 \div 0.01 =$	$0.86 \div 2 =$
$1.5 \div 3 =$	$25 \div 0.5 =$	$1.20 \div 0.3 =$
$0.22 \div 0.02 =$	$15 \div 0.5 =$	$1.8 \div 0.6 =$
$0.54 \div 9 =$	$0.63 \div 0.9 =$	$0.32 \div 0.8 =$
$0.48 \div 2 =$	$2.4 \div 0.4 =$	$10 \div 0.1 =$
$3.7 \div 10 =$	$0.36 \div 9 =$	$0.24 \div 0.1 =$
$4.2 \div 7 =$	$0.5 \div 0.5 =$	$6.3 \div 0.09 =$
$0.18 \div 0.9 =$	$1.2 \div 0.3 =$	$0.6 \div 5 =$

K	A	I	K	S	M	T	E	R	N	K
62	0.12	1.1	100	4	50	0.5	1	0.32	4	4.5
W	P	I	A	C	Q	T	R	I	L	C
3	2.6	0.2	1.9	0.06	60	0.11	1.5	1.75	11	0.99
K	W	M	R	B	A	C	R	M	D	E
0.22	2.4	0.66	0.04	6	3.2	3.5	70	2.5	30	0.9
I	A	L	O	G	L	D	O	A	E	N
0.65	0.43	24	0.7	0.6	32	0.4	0.37	8.1	0.24	9.2

Write the letters in order from the ✱ to the bottom-right corner.

Ongoing Practice

1. Calculate the mean for each data set. Then use the number line to help calculate the MAD. Show your thinking.

a. 6, 6, 7, 7, 9, 12, 12, 13

Mean is

MAD is

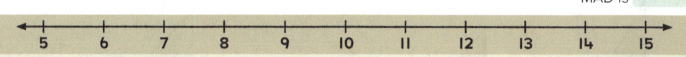

b. 8, 8, 10, 11, 11, 12

Mean is

MAD is

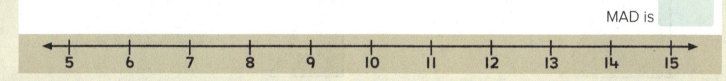

2. Draw and label a tape diagram to show the ratio for each situation.

a. A baker makes fruit and nut muffins. She uses 2 cups of raisins, 5 cups of dried apricots, and 6 cups of mixed nuts.

b. Another baker makes fruit and nut muffins too. He uses 5 cups of raisins, 1 cup of apricots, and 8 cups of mixed nuts.

Preparing for Module 12

Solve each problem. Show your thinking and include the appropriate unit in your answer.

a. Over 6 hours, Dallas spends 30% of her time serving customers and the rest taking inventory. How much time does she spend taking inventory?

b. A farmer sells 25% of his herd of 1,500 cows. How many cows does he have left?

11.7 Ratio: Examining similar triangles

Step In Beatrice is sewing a patchwork cushion cover using large and small triangles. She wants to make a design where the small triangles fit together in exactly the same way as the large triangles.

Which two triangles below should she use to make sure this happens? Explain your choice.

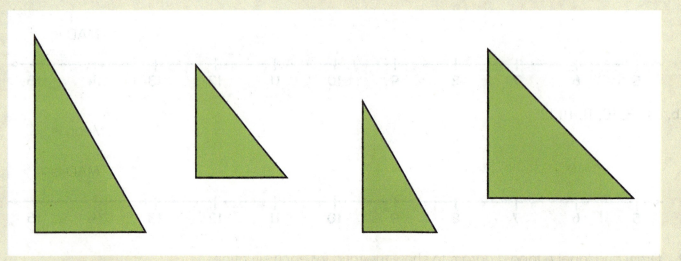

Step Up 1. Cut out all the triangles on the support page your teacher will give you. Then sort them into groups as directed by your teacher.

2. Paste all the triangles that belong in Group A below.

3. Paste all the triangles that belong in Group B below.

4. What do you notice about the angles inside the triangles in Group A and Group B?

Step Ahead Ashley thinks that if you change the length of only one side of a triangle in Group A then only one angle will change too. Daniel thinks all the angles will change. Who is correct? Explain your thinking.

11.8 Ratio: Examining percentage changes of area

Step In   Look at the shape on the right. Estimate its area in cm².

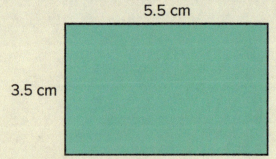

How do you think its area will change if each of the side lengths are doubled?

How will the perimeter change?

I think the area will double if the perimeter is doubled.

I think the area will be more than double if the perimeter is doubled.

Step Up   1. Measure and label each side of the shapes below in centimeters.

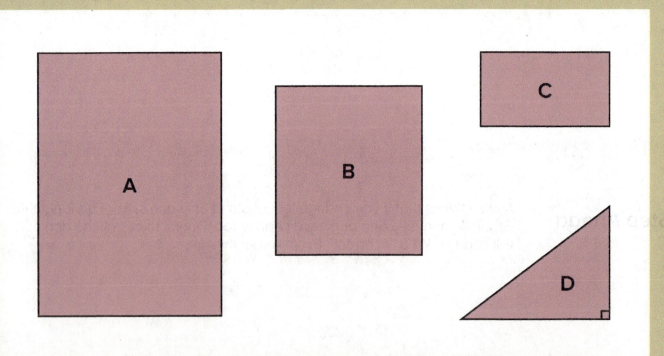

2. Calculate the area of each shape in Question 1. Remember to use the correct unit. If necessary, show your thinking on page 432.

Shape A: ____   Shape B: ____   Shape C: ____   Shape D: ____

3. Redraw each shape from Question 1 so that each side is 50% of the length that it was previously. Label each side of the shapes in centimeters.

4. Calculate the area of each shape in Question 3. Remember to use the correct unit. If necessary, show your thinking on page 432.

Shape A: Shape B: Shape C: Shape D:

5. What do you notice about the areas recorded in Question 2 and Question 4?

Step Ahead

A rectangle has been redrawn so its perimeter is now double what it was originally. If the area of the new rectangle is 64 cm², what could be the length and width of the original rectangle? Show your thinking.

Length: Width:

11.8 Maintaining concepts and skills

Think and Solve

- Aston cycled 15,000 m in 30 minutes.
- Grace cycled 9 km in 15 minutes.
- Tama cycled 30,000 m in 45 minutes.

If they all maintain the same speed, how many kilometers will each cyclist travel in 1 hour?

Aston ☐ Grace ☐ Tama ☐

Words at Work

Write the steps you would use to resize the rectangle below so its area is 40% of the original. Include the length and width of the smaller rectangle and explain your reasoning.

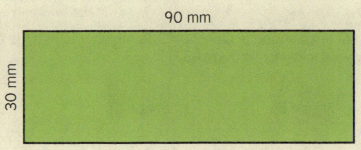

90 mm
30 mm

Ongoing Practice

1. Calculate the quartiles and IQR for the data on this dot plot. Show your thinking.

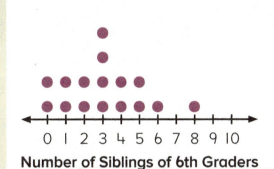

Number of Siblings of 6th Graders

1st quartile	2nd quartile	3rd quartile	IQR

2. Solve each problem. Show your thinking and include the appropriate unit in your answer.

a. Gabriel is making a large batch of granola bars with dried apricots, dates, and mixed nuts in a ratio of 2:3:5. If he uses 3 lb of dried apricots, how many pounds of mixed nuts does he need?

b. Teena has collected a total of 60 stamps. She sorts them into three categories: common, semi-rare, and rare. She figures out the ratio of the categories is 9:5:6. How many of Teena's stamps are rare?

Preparing for Module 12

Use the values to complete the equations. Cross out the values that do not have a match.

a. $5(y + 8) = 100$ $y =$

$5(\square + 8) = 100$

b. $(m - 6)^2 = 9$ $m =$

$(\square - 6)^2 = 9$

c. $t^2 + 15 = 51$ $t =$

$\square^2 + 15 = 51$

d. $2(75 \div b) = 10$ $b =$

$2(75 \div \square) = 10$

Values

6	15
4	12
11	9

11.9 Ratio: Working with percentage changes of area

Step In

Samuru wants to sow grass seed in the backyard to improve the lawn. He measures the yard and finds it is 85 feet long and 57 feet wide. What is an estimate for the total area?

Grass seed is spread at a rate of 2 pounds per 1,000 square feet. About how many pounds of seed are needed?

Natalie suggests that seeding an area half the length and half the width of the backyard would be enough to improve the lawn.

Samuru knows a relationship between changing the dimensions of a rectangle and its area. Without calculating the area again he knows that they will need about $2\frac{1}{2}$ pounds of seed. What do you think he knew that helped him?

Step Up

1. Some facts about Rectangle A are shown in the first table below.

Rectangle	Length	Width	Area (cm²)
A	40	30	1,200

Copies of Rectangle A were made after some changes to its dimensions. Complete each table to show the effects of the changes. Show any calculations on page 432.

a. Keep the width the same but make the length 50% of Rectangle A.

Rectangle	Length	Width	Area (cm²)	Percentage of original area (%)
B				

b. Keep the width the same but make the length 30% of Rectangle A.

Rectangle	Length	Width	Area (cm²)	Percentage of original area (%)
C				

c. Keep the width the same but make the length 10% of Rectangle A.

Rectangle	Length	Width	Area (cm²)	Percentage of original area (%)
D				

2. Different changes were made to Rectangle A in Question 1. Complete each table to show the effects of the changes. Show any calculations on page 432.

   a. Make both the length and width 50% of Rectangle A.

Rectangle	Length	Width	Area (cm²)	Percentage of original area (%)
E				

   b. Make both the length and width 30% of Rectangle A.

Rectangle	Length	Width	Area (cm²)	Percentage of original area (%)
F				

   c. Make both the length and width 10% of Rectangle A.

Rectangle	Length	Width	Area (cm²)	Percentage of original area (%)
G				

3. Describe what you notice about the results for area in Question 1 compared to the results for area in Question 2.

Step Ahead

The total area of the whole square below is 84 m². Complete the sentences. Show your thinking.

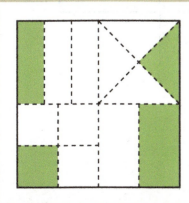

The amount shaded is ____ m².

The shaded area is ____ % of the largest square.

11.10 3D objects: Analyzing pyramid nets

Step In These are nets for pyramids. What do you know about pyramids?

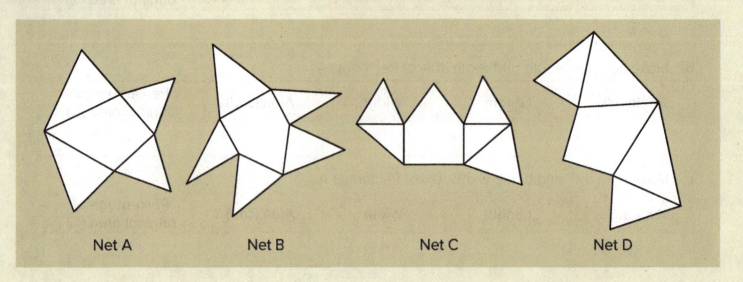

Net A Net B Net C Net D

Color the part of each net that will be the base of the pyramid.

Which nets will create the same object?

Cut out and fold a net from the support page your teacher will give you. Compare your object to that of another student. Write or share what you notice.

How could you calculate the surface area of your pyramid?

Step Up 1. Write why each of these nets will **not** make a pyramid.

a.

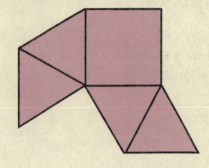

b.

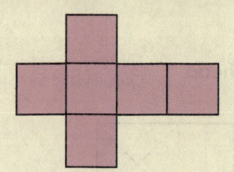

◆ 422

2. Draw a net for the pyramid shown. It does not have to be full size. The base of the pyramid is a rectangle. Then use the dimensions to calculate the surface area. Show your thinking.

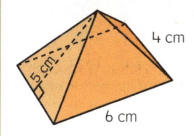

Step Ahead Draw a different net for the pyramid in Question 2.

11.10 Maintaining concepts and skills

Computation Practice

What is a group of tigers called?

★ Use a ruler to draw a line to the correct answer. The line will pass through a letter. Write each letter above its matching answer at the bottom of the page. Some letters are used more than once.

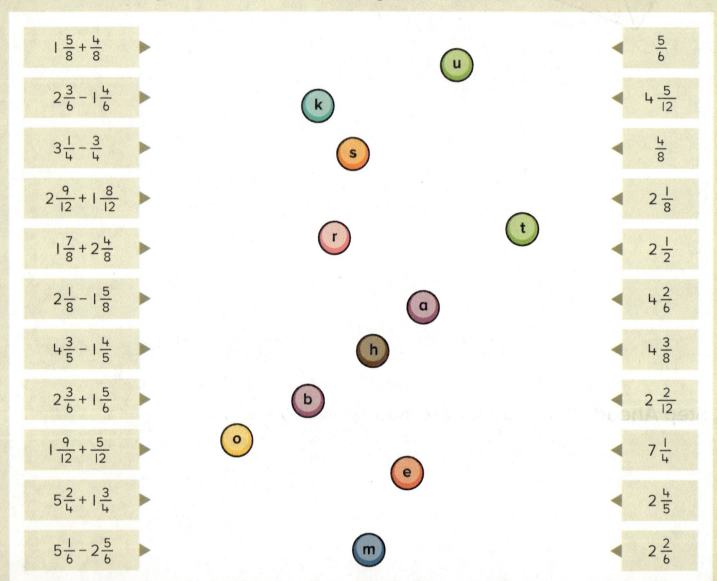

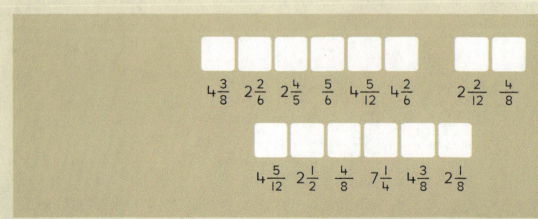

Ongoing Practice

1. A class of Grade 6 students collected data on the number of cousins they each have.

   a. Create a box plot of the data.

10	7	15	18	9	7	6	3	2	4	4	12	1	1	0

   b. Use the box plot to describe the data set on page 432.

2. Measure and label each dimension in millimeters. Then redraw the shape so that each dimension is 30% of the original. Label each new dimension.

Preparing for Module 12

Calculate the value of each variable. Show your thinking.

a. $72 = 3g + 5g$

b. $3f + 6f + 8f = 99 - 48$

c. $2a + 14a = 16 \cdot 4$

d. $48 \div 8 = \frac{1}{2}h + \frac{1}{6}h$

11.11 3D objects: Analyzing prism nets

Step In These are nets for prisms. What do you know about prisms?

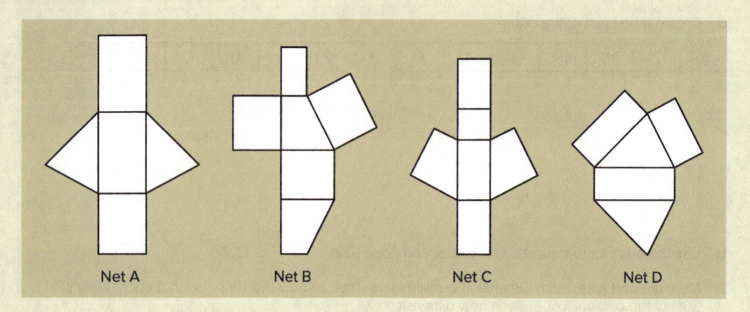

Net A   Net B   Net C   Net D

Color the parts of each net that will be the bases of the prism.
Which nets will create the same object?

Cut out and fold a net from the support page your teacher will give you. Compare your object to that of another student. Write or share what you notice.

How could you calculate the surface area of your prism?

Step Up 1. Write why each of these nets will **not** make a prism.

a.
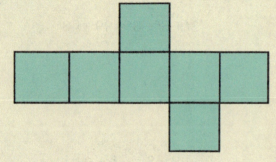

b.
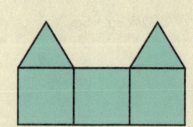

2. Draw a net for the prism below. It does not have to be full-size. Then use the dimensions to calculate the surface area.

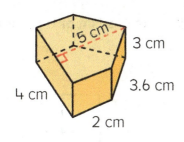

Step Ahead

Draw the extra shape and complete the sentence.

a. Draw one extra 2D shape to make this a **pyramid** net.

This net will make a _____ -based pyramid

b. Draw one extra 2D shape to make this a **prism** net.

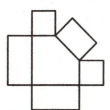

This net will make a _____ -based prism

11.12 3D objects: Creating prism nets

Step In Look at these nets. How do you know if these nets will make a prism or a pyramid?

Color the bases of each prism.
The bases cannot be joined directly together. Why not?

Share how these nets will fold, which faces will share an edge, and which edges will meet at a common vertex.

Step Up 1. Color **red** the nets that will make a prism. Color **blue** the nets that will make a pyramid. Some will not make a prism or pyramid.

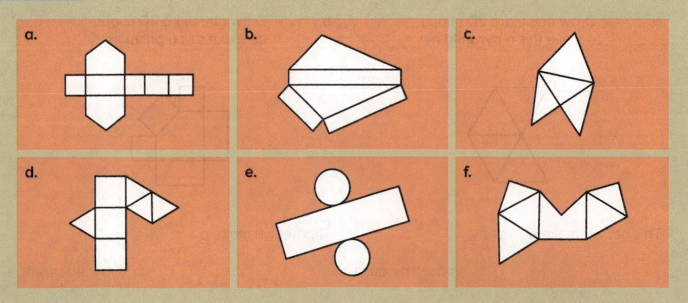

2. Draw a full-size net for this hexagonal-based prism. The bases are regular hexagons so each interior angle is 120°. One base has been drawn for you.

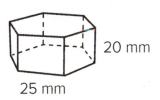

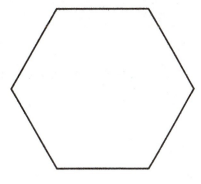

3. Measure and label the dimensions on the net you drew in Question 2. Then calculate the surface area of the prism. Show your thinking on page 432.

Step Ahead Owen wants to make a rectangular-based prism with a surface area as close to 100 cm² as possible. Draw and label the dimensions of a net that matches. The net does not have to be full size.

11.12 Maintaining concepts and skills

Think and Solve

The mean of four numbers is 21.

The range is 15.

The greatest number is 29.

The two middle numbers differ by 3.

Write the numbers in the data set.

☐ ☐ ☐ ☐

What is the median? ☐

Words at Work

Use words from the list to complete true sentences.

a. Ratios between _____ quantities can be written in a _____ way to ratios with two quantities.

b. You can draw a _____ or a _____ to represent a ratio problem.

c. If the length and width of a rectangle are halved then the area of the new rectangle is _____ of the area of the original rectangle.

d. If only the _____ of a rectangle changes by a percentage, then the _____ will change by the same percentage.

e. If all three side lengths of a triangle are doubled in length, then the area of the new triangle will be _____ than the area of the original triangle.

f. A _____ is made by unfolding a 3D object. It is composed of all of the faces of the object.

Word list:
- area
- table
- one-fourth
- four times greater
- net
- similar
- tape diagram
- length or width
- three

Ongoing Practice

1. This table shows the age of some marathon entrants. Use the data to complete the graph.

Marathon Entrants		
Ages	Frequency	Total
18–28	𝍱𝍱𝍱𝍱𝍱𝍱𝍱𝍱 II	42
29–38	𝍱𝍱𝍱𝍱𝍱𝍱𝍱𝍱𝍱𝍱𝍱𝍱 III	58
39–48	𝍱𝍱𝍱𝍱𝍱𝍱𝍱 I	36
49–58	𝍱𝍱𝍱𝍱𝍱	25
59–68	𝍱 III	8
69–78		0

2. a. Measure the dimensions of each shape. Write the letters of the shapes that belong together.

A   B   C   D   E

b. Describe why these rectangles belong together.

Preparing for Module 12

a. Use the rule to complete the table. Make sure the x-coordinates are between 0 and 8.

Rule: $x \div 2 = y$

x (Input)	5			
y (Output)				

b. Graph the coordinates from the table onto the coordinate plane.

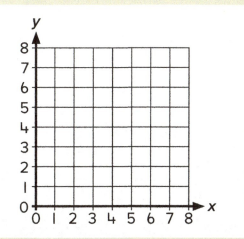

Working Space

12.1 Ratio: Introducing percentages greater than 100%

Step In Ten guests board a boat, and it is now 100% full.

If one more guest were to board the boat, how could you describe the capacity?

How could you use these two squares to show the capacity?

Complete the equivalence statement to show other ways of writing the amount.

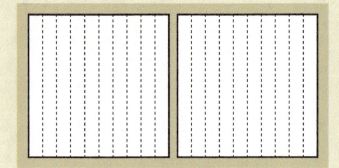

$1\frac{1}{10} = 1\frac{\square}{100} = \frac{\square}{100} = 1.\square = \square\%$

How did you calculate the fraction as a percentage?

Follow the same steps to show one and two-fifths as a percentage.

Complete this equivalence statement to match.

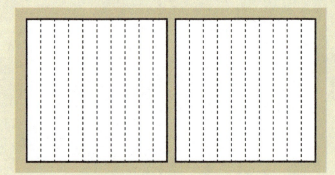

$1\frac{2}{5} = 1\frac{\square}{100} = \frac{\square}{100} = 1.\square = \square\%$

How would you write each percentage as a ratio?

Step Up

1. Each large square represents one whole. Color parts to show each fraction. Write the mixed number or improper fraction, then the decimal fraction and percentage to match.

a. Color $1\frac{1}{4}$

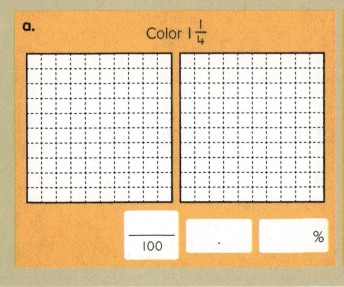

$\frac{\square}{100}$ $\square . \square$ $\square\%$

b. Color $1\frac{3}{4}$

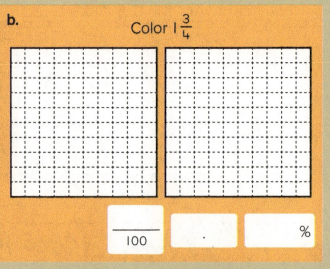

$\frac{\square}{100}$ $\square . \square$ $\square\%$

2. Write the missing numbers in the table to show equivalent amounts.

Common Fraction or Mixed Number (not hundredths)	Common Fraction (hundredths)	Decimal Fraction	Percentage	Ratio
$\frac{3}{4}$				75:100
$\frac{4}{5}$				
	$\frac{170}{100}$			
	$\frac{110}{100}$			
		0.20		
		1.20		
			150%	
			175%	175:100

3. Write some other sets of matching amounts that are **between** 1 and 3.

a.

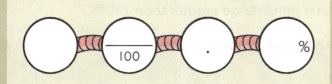

b.

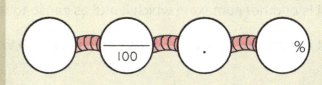

Step Ahead

In a post-match interview, the captain of the Stingrays claimed that his team had given the match 110%. Is it possible to give 110% in effort? Explain your thinking.

12.2 Ratio: Consolidating percentages greater than 100%

Step In

Fiona is sharing the end of month sales report. She says that sales are 150% of last month's sales. What are some possible figures for each month?

Write two possible sales amounts below.

Sales last month	Sales this month

How did you decide what amounts to write?

I started with an easy amount: $80. 100% of $80 is $80, so 50% is $40. So, 150% of $80 is $80 + $40 = $120.

> Finding 150% of an amount means finding 100%, then 50% more.
>
> Finding an increase of 150% means adding 100% of the amount to itself then adding 50% more.

I did it a different way. I knew that 10% of $80 is $8. I want to find 150%, so that's 15 • $8.

What is another context in which it makes sense to use a percentage greater than 100?

What is a context in which it does not make sense to use a percentage greater than 100?

Step Up

1. Read each situation. Color the situations where it makes sense to have a percentage greater than 100%.

a.	A car manufacturer reports that the new engine is 120% of the size of the old engine.	b.	If compared to an old USB, the storage capacity of the new USB has increased by 200%.
c.	Allan's inbox for his email account is now 124% full.	d.	107% of voters approved the new licensing laws.
e.	106% of the grade level passed the end of year examination.	f.	An airline has a policy of selling 110% of seats for each flight.

2. Complete the missing parts to show equivalent amounts.

a.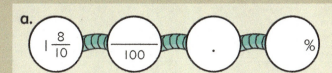

b.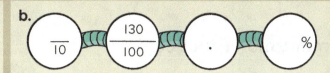

c.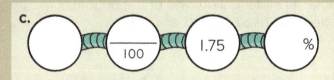

d.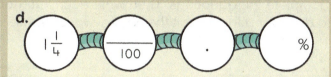

e.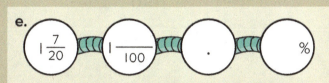

f.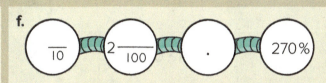

3. Write two ratios to match each percentage.

a. 140% = 14:10 =	b. 200% = =	c. 150% = =
d. 250% = =	e. 120% = =	f. 110% = =

4. Solve each problem. Show your thinking. Remember to include the appropriate unit in your answer.

a. An engineer claims that the battery life of their new cell phone is 200% longer than their old cell phone. The battery in their old cell phone would last for about 24 hours. What is the battery life for the new cell phone?

b. The school fundraising committee reports that their end of year profits are 175% of the previous year. Last year they raised $1,000. What amount did they raise this year?

Step Ahead Charlotte says that these two amounts are equivalent. Do you support her statement? Show your thinking.

$105\% = \frac{21}{20}$

12.2 Maintaining concepts and skills

Computation Practice

★ Complete the equations. Then write each letter above its matching answer at the bottom of the page to discover a fact about the natural world. Some letters are used more than once.

8 • 3.2 = 25.6	e	1.5 • 9 = 13.5	h
4 • 0.17 = 0.68	s	0.4 • 0.5 = 0.2	o
0.6 • 0.9 = 0.54	t	2.4 • 8 = 19.2	a
5.3 • 0.8 = 4.24	w	6 • 0.48 = 2.88	n
5 • 6.2 = 31	i	1.2 • 7 = 8.4	r
0.07 • 12 = 0.84	f	4.8 • 3 = 14.4	u
0.8 • 0.7 = 0.56	l	11 • 0.04 = 0.44	v
2.2 • 0.5 = 1.1	y		

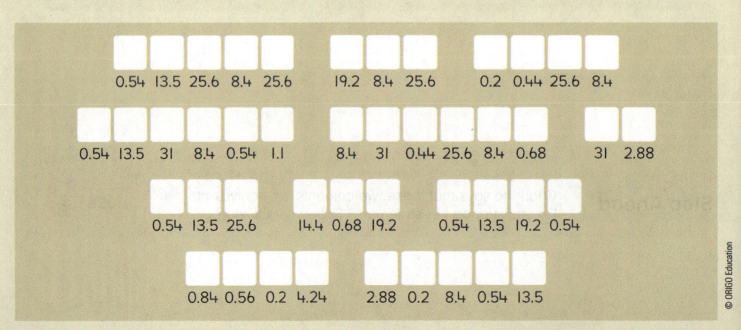

Ongoing Practice

1. a. Measure the dimensions of each shape with a ruler marked in millimeters. Write the letters of the shapes that belong together.

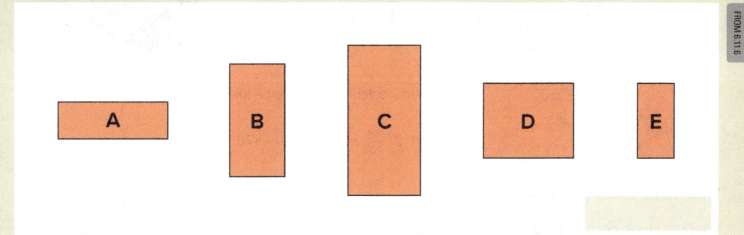

b. Describe why these rectangles belong together.

2. Write the missing numbers in the table to show equivalent amounts.

Common Fraction or Mixed Number (not hundredths)	Common Fraction (hundredths)	Decimal Fraction	Percentage	Ratio
	$\frac{50}{100}$			50:100
$1\frac{3}{4}$		1.75		
	$\frac{80}{100}$			80:100
$1\frac{3}{10}$			130%	

Preparing for Next Year

Interpret each expression. Then write the number that is **opposite**. Use the number line to help your thinking.

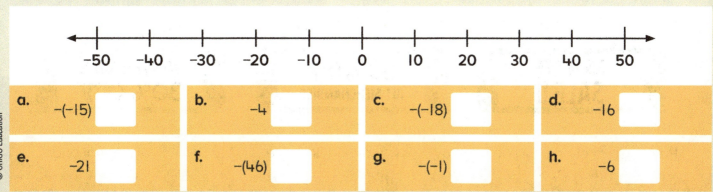

a. −(−15)
b. −4
c. −(−18)
d. −16
e. −21
f. −(−46)
g. −(−1)
h. −6

12.3 Ratio: Using complementary percentages

Step In

This cell phone is on sale.

How would you calculate the new price of the cell phone?

Cathy calculates the price like this.

$$\frac{15}{100} \cdot \frac{200}{1} = \frac{3000}{100}$$

$$\frac{3000}{100} = \frac{30}{1} = 30$$

$$200 - 30 = 170$$

Hugo calculates the price like this.

$$\frac{85}{100} \cdot \frac{200}{1} = \frac{1700}{100} = 170$$

What steps does each person follow?

Why did Hugo calculate 85% of $200 to find his answer?

Can you think of another way to calculate the price of the new cell phone?

How would you calculate the new price of a cell phone that costs 10% less than $80?

I know that 10% less means I pay 90% of the original price. 90% of $80 is equivalent to $\frac{9}{10} \cdot 80 = \frac{720}{10} = 72$.

Step Up

1. Look at each sign. Then write the percentage of the total that needs to be paid.

a. **10% OFF ALL PRODUCTS**
Pay ☐ % of original price

b. **20% OFF ALL MERCHANDISE**
Pay ☐ % of original price

c. **CYBER MONDAY SALE 75% OFF**
Pay ☐ % of original price

d. **25% OFF SALE !!**
Pay ☐ % of original price

e. **60% OFF ALL MERCHANDISE**
Pay ☐ % of original price

f. **WINTER SALE 30% OFF**
Pay ☐ % of original price

2. Use Hugo's method to calculate the amount that you would pay.

a.

b.

c.

d.

e.

f.

g.

h.

Step Ahead

Jennifer wants to buy a new television. She finds the same television at two different stores. The price at Store A is $1,500 with 15% off. The price at Store B is $1,400 with 5% off. Which store has the less expensive option? Show your thinking.

12.4 Ratio: Using percentages greater than 100%

Step In

A store sells shirts for 50% more than what they pay to buy them.

That's 150% of the amount that they paid.

So a shirt that they buy for $10 will be sold for $15.

The store buys a shirt for $20. What will be the selling price?

Cary calculates the amount like this.

100% of 20 = 1 • 20
so
150% of 20 = 1.5 • 20
1.5 • 20 = (1 • 20) + (0.5 • 20) = 30

Emily calculates the amount like this.

$$\frac{150}{100} \cdot 20 = \frac{3000}{100}$$
$$\frac{300}{10} = \frac{30}{1} = 30$$

Converting percentages to common fractions or decimal fractions can help you find a solution.

What steps does each person follow?

Why does Cary multiply 20 by 1.5?

How could you use each strategy to calculate the selling price of a shirt that costs $50 to buy?

Step Up

1. Use Cary's or Emily's method to calculate each selling price. Show your thinking.

a.  Buy for: $25

Sell for 20% more

b.  Buy for: $50

Sell for 50% more

c.  Buy for: $30,000

Sell for 25% more

d.  Buy for: $2

Sell for 40% more

2. Use Cary's or Emily's method to calculate each retail price. Show your thinking.

a. Buy for $75   Sell for 10% more

b. Buy for $30   Sell for 50% more

c. Buy for $1.50   Sell for 100% more

d. Buy for $20   Sell for 120% more

3. Compare the two prices. Then write the selling price as a percentage of the buying price. Show your thinking.

a.

b.

Step Ahead

A manufacturer makes a designer jacket for $45. They sell it to a supplier that increases the price by 10%. The supplier sells the jacket to a store that increases the price by 100%. What is the final selling price of the jacket? Show your thinking.

12.4 Maintaining concepts and skills

Think and Solve

a. What percentage of all of the square numbers from 1 to 100 are even? ▢

b. Write how you figured it out.

Words at Work

Write in words how you solve this problem.

A farmer feeds his chickens a bag of feed per day. He is concerned that they are not getting enough food so he increases the amount by 50%. After 4 weeks, he still has some concerns and the local vet advises him to increase what he is currently feeding them by 150%. A bag of feed costs $5.00. How much does it cost the farmer to feed his chickens each week?

Ongoing Practice

1. a. Measure the dimensions of each triangle with a ruler marked in millimeters. Write the letters of the shapes that belong together.

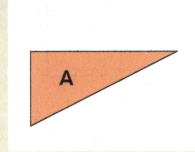 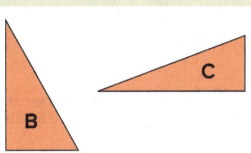 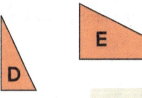

   b. Describe why these triangles belong together.

2. Write two ratios to match each percentage.

   a. 150% = ___ = ___

   b. 120% = ___ = ___

   c. 160% = ___ = ___

   d. 110% = ___ = ___

   e. 130% = ___ = ___

   f. 140% = ___ = ___

Preparing for Next Year

a. Write each situation as a rate. You will need to change one of the quantities to 1.

Mateo bought 6 plants for $15.	A plumber charges $260 for 4 hours of work.
The rate is ___.	The rate is ___.

b. Use the stories above to answer the questions below. Show your thinking.

How many plants could Mateo buy with $65?	How much would the plumber charge for 3 hours of work?

ORIGO Stepping Stones · Grade 6 · 12.4

445

12.5 Algebra: Simplifying expressions involving percentages

Step In

$\frac{3}{5}$ of a quart of punch is left in a pitcher. 20% of it is apple juice. What fraction of a quart is the apple juice?

How would you solve this problem?

Kayla solves the problem like this.

$$20\% \text{ of } \frac{3}{5}$$
$$= \frac{20}{100} \cdot \frac{3}{5}$$
$$= \frac{60}{500}$$
$$= \frac{6}{50}$$
$$= \frac{3}{25} \text{ qt}$$

*Remember that **of** means to multiply.*

Converting between common fractions and percentages could make the problem easier to solve.

What steps did she follow?

What common fractions could Kayla use instead of $\frac{20}{100}$?

Would any of them make the calculations easier?

A 5 oz jar of green paint is made using 90% yellow paint and 10% blue paint. Ethan has plenty of yellow paint, but only 4 oz of blue paint.

How many 5 oz jars of green paint can he make?

Try drawing a picture if you're having trouble understanding a problem.

Green Paint 5 oz

🔍 **Mixture**
10% blue
90% yellow

Step Up

1. Read the problem. Circle each expression that matches a method to solve it. There is more than one method.

A runner has completed $\frac{1}{5}$ of a 1-mile race. 30% of the remaining race is uphill. How much uphill running remains for the runner?

a.	b.	c.	d.	e.
70% of $\frac{1}{5}$	30% of $\frac{4}{5}$	80% of 30%	0.3 • 0.8	30% of 20%

446

2. Use two of the methods you circled in Question 1 to calculate the distance that is run uphill. Show your thinking and remember to include the appropriate unit in your answer.

Method 1:

Method 2:

3. Simplify each expression. Show your thinking.

| a. 5% of 15 | b. 12% of $\frac{1}{2}$ | c. 3 ÷ (10% of 90) |
| d. $2(\frac{3}{5}$ of 60%) | e. 25% of 0.2 | f. 15% of $\frac{90}{100}$ |

Step Ahead

Harvey makes 20 oz of brown paint by mixing blue, red, and yellow. The ratio of B:R:Y is 20:30:30. Show how you solve the problems below.

a. How much yellow paint did he use?

b. What percentage is the yellow paint of the whole 20 oz?

12.6 Algebra: Solving word problems involving percentages

Step In

Carmen works at an animal shelter. She found homes for 200 cats and dogs last year. 90 of the adopted animals were cats.

What percentage were dogs?

How would you solve this problem?

Juan solves the problem like this.

$(200 - 90) \div 200$
$= \frac{110}{200}$
$= \frac{55}{100}$
$= 55\%$

Percent means "per hundred." I can halve the number of animals to work with one hundred.

I'll also need to halve the number of cats. This keeps the relationship between the two amounts the same.

Describe the steps that Juan follows.

Why did he halve the number of cats and the total number of animals?

What other word problem could you write to match the expression Juan wrote?

Step Up

1. Read the word problem. Then circle each expression that shows how to solve it. There is more than one option.

   A store buys a rare baseball card for $30. They then sell the card to a customer for 20% more than what they paid. What was the amount the customer paid?

a.	b.	c.	d.	e.
$30 + (20\%$ of $30)$	$30 \cdot 0.2 + 30$	$(\frac{2}{10} \cdot 30) + 30$	$(30 + 0.2) \cdot 30$	$30 + 20 \cdot 30$

2. Calculate the selling price of the card using two different methods. Show your thinking.

   Method 1:

   Method 2:

3. Solve each problem. Show your thinking and include the appropriate unit in your answer.

a. Ringo wants to buy a game that costs $30. The game store is having a 15% off sale. As a member of the store, Ringo is given another 5% off the original price. What amount does he save?

b. There are 60 animals at the animal shelter. $\frac{3}{4}$ of the animals are cats. Of those cats 40% are female. How many male cats are at the animal shelter?

c. Twenty-four students order 12 large pizzas to share equally. Each student eats 40% of their share of pizza. What percentage of pizza remains?

d. Jacinta's fish tank holds 40 gallons of water. It is $\frac{2}{3}$ full. The instructions say that the tank should be at least 80% full. How many gallons of water need to be added?

Step Ahead Solve this problem. Show your thinking.

Corey's phone has only 30% battery power left. His phone has been on for 8 hours. If his battery continues to lose power at the same rate, about how many hours of battery life does he have left?

12.6 Maintaining concepts and skills

Computation Practice — Who invented the rubber balloon?

★ Look at the relationship between the dividend and the divisor to decide what strategy to use to solve each equation. Then find each quotient in the grid below and cross out the letter above. Then write the remaining letters at the bottom of the page.

$\frac{16}{20} \div \frac{4}{20} =$ ☐

$\frac{3}{4} \div \frac{1}{4} =$	$8 \div \frac{1}{6} =$	$\frac{6}{9} \div \frac{2}{3} =$	$\frac{8}{9} \div \frac{1}{9} =$	$\frac{5}{12} \div \frac{3}{5} =$
$\frac{1}{6} \div \frac{2}{5} =$	$\frac{5}{7} \div \frac{4}{3} =$	$\frac{1}{3} \div \frac{3}{4} =$	$\frac{3}{8} \div \frac{3}{4} =$	$\frac{12}{15} \div \frac{2}{15} =$
$\frac{3}{5} \div \frac{2}{7} =$	$9 \div \frac{4}{9} =$	$6 \div \frac{2}{3} =$	$\frac{1}{2} \div \frac{2}{3} =$	$\frac{10}{6} \div \frac{1}{3} =$
$\frac{10}{6} \div \frac{1}{6} =$	$\frac{3}{5} \div \frac{2}{7} =$	$\frac{3}{4} \div \frac{3}{8} =$	$\frac{5}{4} \div 5 =$	$10 \div \frac{2}{5} =$
$\frac{3}{7} \div \frac{1}{3} =$	$4 \div \frac{1}{3} =$	$\frac{1}{4} \div \frac{3}{4} =$	$5 \div \frac{1}{4} =$	$\frac{1}{2} \div \frac{1}{14} =$

H	M	P	K	I	M	T	E	C	X
$\frac{1}{2}$	$\frac{5}{6}$	$\frac{3}{4}$	25	11	$\frac{4}{9}$	4	$2\frac{1}{10}$	$6\frac{1}{2}$	7
N	H	I	A	E	Q	U	A	I	E
$1\frac{2}{7}$	$\frac{7}{12}$	2	$\frac{5}{6}$	10	20	$\frac{5}{12}$	8	48	$2\frac{3}{7}$
B	L	A	F	B	L	A	B	N	R
9	$1\frac{1}{3}$	6	$\frac{3}{10}$	$2\frac{1}{7}$	3	$\frac{12}{13}$	5	12	$\frac{2}{3}$
J	A	I	D	G	A	M	O	Y	F
1	$\frac{3}{8}$	$\frac{15}{28}$	$\frac{7}{9}$	$\frac{1}{3}$	$\frac{2}{5}$	$\frac{25}{36}$	$\frac{1}{4}$	$\frac{2}{7}$	$20\frac{1}{4}$

Write the letters in order from the ✱ to the bottom-right corner.

☐☐☐☐☐☐   ☐☐☐☐☐☐☐

Ongoing Practice

1. a. Measure and label each side of the shape below in centimeters. Then write the area.

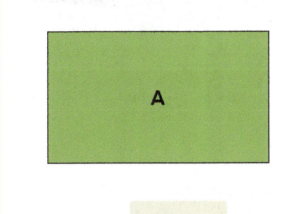

Area _____   Area _____

b. Redraw Shape A so that each side is 50% of the length that is was previously. Label each side of the new shape (B) in centimeters. Then calculate the area.

c. What do you notice about the area of Shape A and the area of Shape B?

_____

2. Calculate the amount you would pay for each. Show your thinking.

a.

b.

Preparing for Next Year

This is a net of a prism that has an equilateral triangular base. Use the same color to show the parts that have the same area. Then calculate the surface area of the object. Show your thinking. Remember to include the unit in your answer.

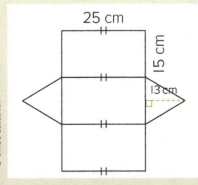

12.7 Algebra: Solving equations with percentages and variables

Step In

Blake bought some books at a 20% off sale. He saved a total of $6.

Which equation below matches the original price of the books?

Let b represent the original price.

$b = \frac{80}{100} \cdot 6$ $20\% \cdot b = 6$ $6 = b - (\frac{20}{100} \cdot b)$

Two friends agree to use the middle equation.

They use slightly different methods to calculate the original price of the books.

Kylie solves it like this.

$20\% \cdot b = 6$
$\frac{1}{5} \cdot b = 6$
$b \cdot \frac{1}{5} = 6$
$b \cdot \frac{1}{5} \cdot 5 = 6 \cdot 5$
$b = 30$

Kevin solves it like this.

$20\% \cdot b = 6$
$0.2 \cdot b = 6$
$b \cdot 0.2 = 6$
$b \cdot 0.2 \div 0.2 = 6 \div 0.2$
$b = 30$

Even though they do not mean the same thing in context, $\frac{1}{5} \cdot b$ has the same product as $b \cdot \frac{1}{5}$.

What steps did each person follow?

Why did Kylie multiply both sides of her equation by 5?

Why did Kevin divide both sides of his equation by 0.2?

If I replace the variable with my solution, I can check if my answer is correct.

How can you check that the answer is correct?

Step Up

1. Calculate the value of the variable. Show your thinking.

a. $10\% \cdot h = \$8$

b. $4 = 25\% \cdot j$

2. Solve each of these. Show your thinking.

a. $20\% \cdot h = 12$

b. $10 = 25\% \cdot p$

c. $7 = 70\% \cdot c$

d. $40\% \cdot a = 16$

3. Check each answer and cross out each incorrect answer. Show your thinking.

a. $30\% \cdot p = \$6$ $p = \$21$

b. $x = 30 + (20\% \cdot 30)$ $x = \$36$

c. $25\% \cdot z = 9$ $z = 13$

d. $28 = 50\% \cdot y$ $y = 13$

Step Ahead

A cell phone is selling for $300. The salesperson takes 20% off the selling price. Calculate the new selling price using each method. Let c represent the new price. Then circle your preferred method.

a. $c = 300 - (0.2 \cdot 300)$

b. $c = 300 - (\frac{20}{100} \cdot \frac{300}{1})$

c. $c = 300(1 - 0.2)$

ORIGO Stepping Stones · Grade 6 · 12.7

12.8 Algebra: Solving word problems with percentages and variables

Step In

After each birthday, Manuel gets 10% more weekly allowance than the year before. This year he gets $2 more allowance than last year.
How much weekly allowance was Manuel earning last year?

What equation could you write to match this problem?

Michelle represents the problem like this.

$10\% \cdot a = \$2$

$2 is 10% of the amount that Manuel got when he was one year younger. I wonder what he will get next year...

What does a represent in Michelle's equation?

How does the equation match the word problem?

How would you calculate the allowance that Manuel earns this year?

Write another word problem to match the equation that Michelle wrote.

Step Up

1. Read each word problem. Circle the equation that matches the problem. Then calculate the value of the variable. Show your thinking.

   a. Fatima saves her spare change and $1 bills. At the end of the year she has $100 in coins. This represents 25% of her total savings. How much money has she saved? Let m represent her total savings.

   $25\% \cdot \$100 = m$ $25\% \cdot m = \$100$

   b. Koda has a part-time job. His boss gives him a bonus of $400. This is 20% of his total earnings for the year. How much did Koda earn before his bonus? Let e represent the unknown amount.

   $\$400 = 20\% \cdot e$ $\$400 \cdot 20\% = e$

2. Solve each problem. Show your thinking.

a. A bookstore reports $15,000 in sales for December. This is 20% of the previous 11 months in sales combined. What is the total amount of sales for the first 11 months of the year? Let b represent the unknown amount.

b =

b. Helen owes some money to her brother. She pays him back at $15 a week, which is 30% of the total amount that she owes. How much does Helen owe her brother? Let d represent the unknown amount.

d =

c. The number of model cars in Deon's collection increased by 5% from the previous year. This year he bought 12 model cars. How many cars were in Deon's collection the previous year? Let c represent the unknown amount.

c =

d. Hannah is sharing the sales report for February. She says that the sales are only 40% of last month's sales. She reports $2,000 in sales for February. What are the total sales for January? Let p represent the unknown amount.

p =

3. Write a word problem to match the equation. Then find the value of the variable. Show your thinking on page 470.

$90 = 60\% \cdot y$ y =

Step Ahead

Megan says that g is greater than $20. Hunter disagrees. He says that g is less than $20. Explain who you agree with. You can use page 470 if you need more space. Do not calculate the exact answer.

$20\% \cdot g = \$3.95$

12.8 Maintaining concepts and skills

Think and Solve

Use the clues and the graph to figure out the answers.

Clues

- There are $1\frac{1}{2}$ times as many girls in Grade 7 as in Grade 6.
- The average number of girls in each grade is 30.
- The average number of boys in each grade is 34.

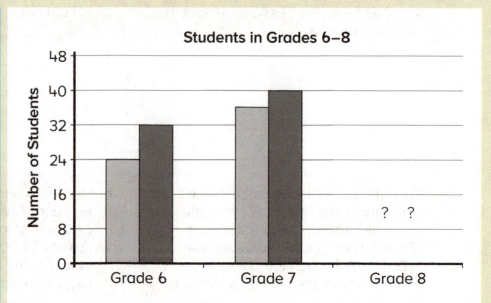

a. Do the dark gray bars represent boys or girls?

b. How many boys and girls are in Grade 8?

Words at Work

Research and write about where percentages greater than 100 can be found in everyday life.

Ongoing Practice

1. Some facts about Rectangle A are shown in the first table below.

Rectangle	Length	Width	Area (cm²)
A	50	20	1,000

Copies of Rectangle A were made after some changes to its dimensions. Complete each table to show the effects of the changes. Show your thinking on page 470.

a. Keep the width the same but make the length 20% of Rectangle A.

Rectangle	Length	Width	Area (cm²)	Percentage of original area (%)
B				

b. Keep the width the same but make the length 50% of Rectangle A.

Rectangle	Length	Width	Area (cm²)	Percentage of original area (%)
C				

2. Simplify each expression. Show your thinking.

a. 40% of $\frac{3}{5}$

b. $5 \div (5\% \text{ of } 40)$

c. 45% of 90%

Preparing for Next Year

Order the data from least to greatest. Write the first, second, and third quartile. Then calculate the IQR.

a. 20, 15, 17, 12, 22, 17, 15, 15, 22, 14

1st quartile ☐   2nd quartile ☐   3rd quartile ☐   IQR ☐

b. 3, 4, 7, 2, 2, 1, 6, 5,

1st quartile ☐   2nd quartile ☐   3rd quartile ☐   IQR ☐

ORIGO Stepping Stones · Grade 6 · 12.8

12.9 Algebra: Generating and graphing variables

Step In

A car is traveling at 60 miles per hour down a straight highway. It stays at that speed for half an hour. Which graph below represents the distance traveled during that time?

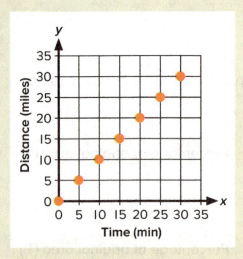 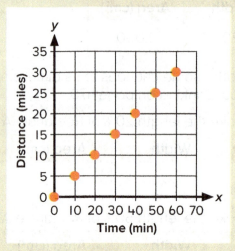 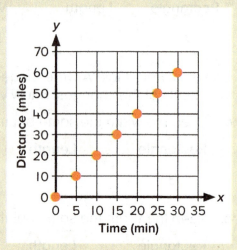

What information helped you decide?
How could you determine the distance traveled in 16 minutes? 48 minutes?

How would the graphs change if the car traveled at 40 miles per hour?
What if the car started from 0 miles per hour?

What are the two variables in the story?
Which variable do you think is independent? Why?

I don't think the graph would make a straight line until the car reached 60 miles per hour. I think that would happen pretty quickly though.

Step Up

1. Your teacher will give you materials to complete an activity.

   a. Write the name of the activity:

   b. What are the variables in your activity?

   c. Create a table to record your data below.

2. Which variable is the independent variable in your activity? Explain your answer.

3. Graph your data from page 458. Choose a suitable range for each axis and label them. Remember to put the independent variable on the x-axis.

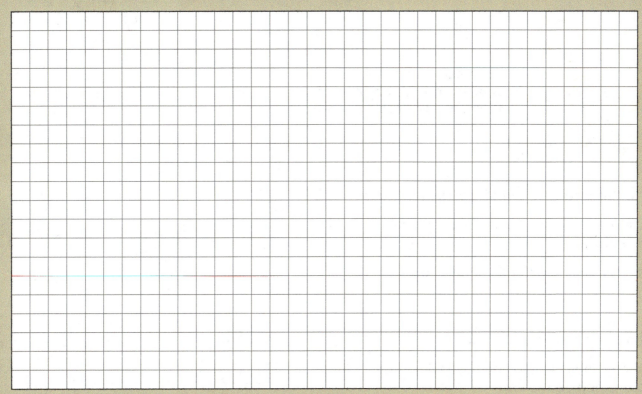

4. Write an equation that describes the relationship between the variables. Use page 470 to describe what each variable represents.

5. Write a rate that describes the relationship between the variables.

Step Ahead

A Boeing 747 airplane at 35,000 feet can fly at about 560 miles per hour. About how long has it been flying if it has covered 60 miles at that speed?

Algebra: Generating and graphing variables (non-equivalent ratios)

Step In

Some tables and chairs are being arranged in a conference room. The picture on the right shows a top view of the furniture.

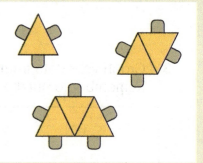

What relationship do you see between the number of chairs and tables?

> I can see that the number of chairs increases by one each time an extra table is added.

Look at this arrangement of tables and chairs.

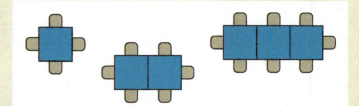

What relationship do you see between the tables and chairs?

How is this similar to the previous example?

Look at these tables and chairs.

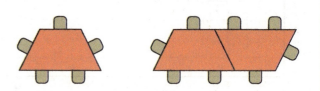

How many chairs would you have with three tables that are joined end to end?

Step Up

1. Write the missing numbers to record the number of chairs needed for each type of table. You can use page 470 to show your thinking. The first few have been done for you.

Number of tables	1	2	3	5	10
△ Number of chairs	3	4			
▢ Number of chairs	4				
⏢ Number of chairs	5				

2. Look at the results from Question 1. Which is the dependent variable, the number of tables or the number of chairs? Explain your thinking.

460

3. Write the results from Question 1 as ordered pairs in the table on the right. The first few have been done for you.

Tables and Chairs						
△	(1, 3)	(2, 4)				
■	(1, 4)					
⬢	(1, 5)					

4. a. Plot each set of ordered pairs from Question 3 onto the graph on the right. Use the colors below to show the different types of table.

b. On the graph, join dots of the same color with a straight line.

5. Can you use the graph to determine the number of tables required for any number of chairs? Why?

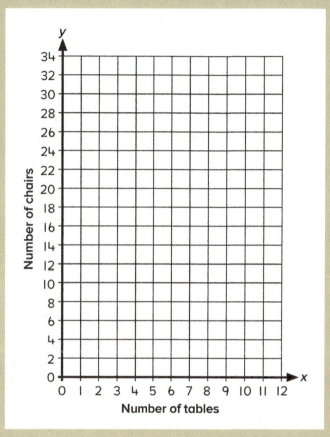

6. Write what you notice about the ratios between the tables and chairs.

Step Ahead Write an equation that gives the rule for calculating the number of chairs (c) for any number of square tables (t) joined end-to-end. Show your thinking on page 470.

12.10 Maintaining concepts and skills

Computation Practice

★ Use a ruler to draw a line to connect expressions that have the same product. Each line will pass through a letter. Write each letter above its matching product at the bottom of the page to reveal a historical fact. Some products appear more than once.

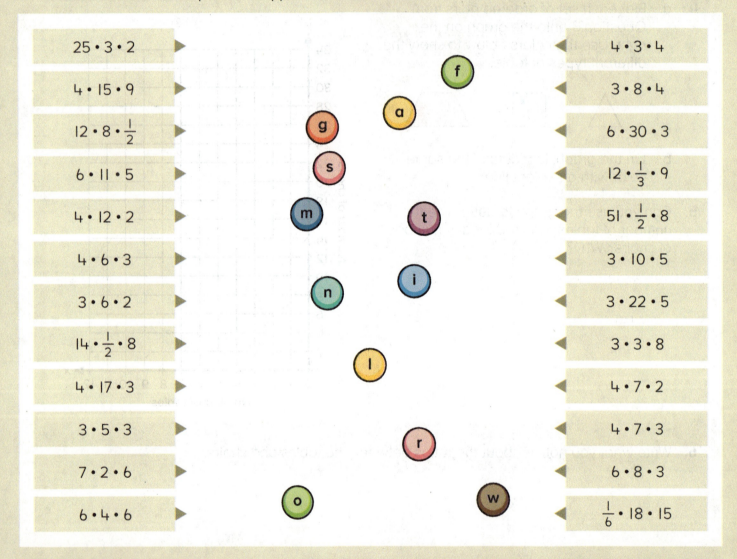

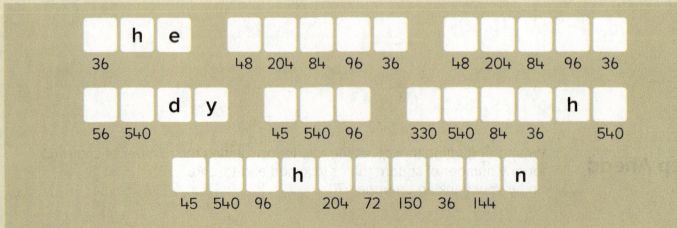

Ongoing Practice

1. Draw a net for the prism below. It does not have to be full-size. The bases are equilateral triangles. Then use the dimensions to calculate the surface area.

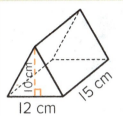

2. Solve each problem. Show your thinking and include the appropriate unit in your answer.

   a. Andrea wants to buy a guitar that costs $150. The music store has a 25% off sale. When she buys the guitar she is given an extra 5% off. What amount does she pay for the guitar?

   b. A baker made 45 muffins to sell. Two-thirds of them were banana muffins. 30% of the banana muffins also had walnuts in them. How many banana muffins do not have walnuts?

Preparing for Next Year

Solve each problem. Show your thinking and include the appropriate unit in your answers.

   a. Carter mixes pure essential oils to make air fresheners. In one batch he mixes teaspoons of orange, cypress, and sage oils in the ratio of 3:5:7. If he uses 84 teaspoons of sage oil, how much orange and cypress oil will he use in total?

   b. Monique has cows, horses, and chickens on her farm in the ratio of 6:2:15. She has 30 cows. How many chickens does she have?

Algebra: Generating and graphing variables (approximate ratios)

Step In

A zoo has a display that shows the mass of their mother animals and newborn babies.

Mother: 44 oz
Baby: 1.2 oz

Mother: 8,930 lb
Baby: 205 lb

Mother: 243 lb
Baby: 1.9 lb

Mother: 185 lb
Baby: 1.2 oz

Look at the ratio between the mother's mass and the baby's mass for each type of animal. What whole number ratio can you write for each type of animal so that the baby's mass is represented as 1?

Rabbit   ____ : 1

Elephant   ____ : 1

Tiger   ____ : 1

Panda   ____ : 1

I'll have to write approximate ratios because I won't be able to use whole numbers if the mass of the baby must be shown as 1.

Not all mother and baby animals have exactly the same mass as the ones shown above but the ratios between them will be about the same. What are some other possible masses for a mother elephant and a newborn elephant so the ratio between them is about the same as above?

Step Up

1. Work with a partner to measure your height then foot length to the nearest half-inch. Record this data as an ordered pair. ____

2. Work with your teacher to collect the same measurement data for all students in your class. Record the measurements as ordered pairs in each cell of the table below. Remember to include your own.

3. Graph each ordered pair you collected in Question 2. Choose a suitable range for each axis and label them.

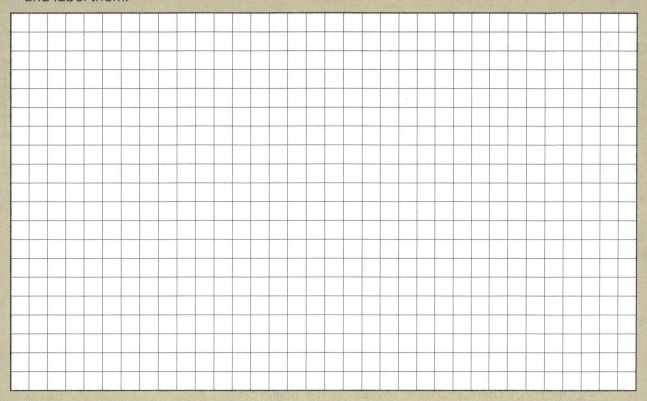

4. What rate would you use to describe the relationship between height and foot length? Explain your thinking.

5. Write an equation that describes the relationship between height (h) and foot length (f).

6. What foot lengths are expected for the following heights?
   a. 71 in
   b. 38 in

7. What heights are expected for the following foot lengths?
   a. 4 in
   b. 14 in

Step Ahead The mass of a mother giraffe is about 10 times greater than its baby. The mass of a mother gorilla is about 50 times greater than its baby. The ratio between the mass of a baby gorilla and the mass of a baby giraffe is about 1:60. If a mother giraffe weighs 2,248 lb, about how much does a mother gorilla weigh? Show your thinking on page 470.

Algebra: Generating and graphing variables (non-linear)

Step In

A courier driver kept track of the cumulative distance covered during one work shift. The results are shown on the right.

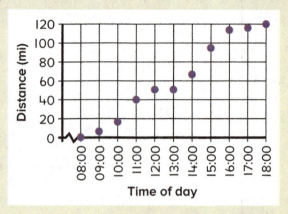

What do you notice about the times written on the *x*-axis?

> 24-hour time can be used to avoid using a.m. and p.m.

What can you say about the distance traveled in each hour? Is it always the same? Why do you think it changes?

Why does the graph only go upward and never downward?

Which variable do you think is independent?

> I think distance is affected by time, but not evenly. Other things can affect the distance traveled by the driver.

Step Up

The temperatures over a day in Omaha, Nebraska, are shown below.

Time	Temp. (°F)	Time	Temp. (°F)	Time	Temp. (°F)
00:00	47	08:00	35	16:00	56
01:00	45	09:00	36	17:00	58
02:00	43	10:00	37	18:00	59
03:00	41	11:00	39	19:00	58
04:00	39	12:00	41	20:00	54
05:00	37	13:00	43	21:00	51
06:00	36	14:00	48	22:00	49
07:00	34	15:00	54	23:00	46

1. Which variable is the independent variable? Explain your thinking.

2. Graph each pair of times and temperatures from page 466. Choose a suitable range for each axis and label them. Remember to put the independent variable on the x-axis.

3. Is there a ratio or equation that you can write to describe the relationship between time and temperature for any time of day? Explain your thinking.

Step Ahead

A few days after the temperatures on page 466 were recorded, these temperatures were also recorded for Omaha.

What are some main differences between these temperatures compared to the temperatures on the previous page? What might have caused these differences? Explain your thinking on page 470.

Time	Temp. (°F)
00:00	45
03:00	36
06:00	37
09:00	39
12:00	41
15:00	43
18:00	48
21:00	54

12.12 Maintaining concepts and skills

Think and Solve

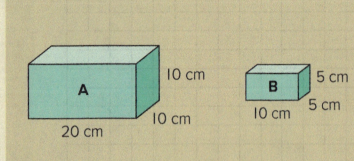

a. What ratio describes the relationship between the surface area of Prism A and the surface area of Prism B?

b. What ratio describes the relationship between the volume of Prism A and the volume of Prism B?

Words at Work

Write the answer for each clue in the grid. Use words from the list. Some words are not used.

Clues Across

2. The expressions 30% of $\frac{4}{5}$, 80% of 30%, and 0.3 • 0.8 all have the same ___.

7. The x-axis shows the ___ variable.

Clues Down

1. When solving an equation involving a variable, you can check your answer by replacing the variable with your ___.

3. The ratios 4:5 and 20:25 are ___.

4. $\frac{1}{5}$ • b has the same ___ as b • $\frac{1}{5}$.

5. In the expression 20% of 30, *of* means to ___.

6. Percent means per ___.

divide	multiply	value	subtract
hundred	percentage	product	quotient
equal	equivalent	answer	solution
independent	dependent	same	thousand

468

Ongoing Practice

1. Color red each net that will make a prism. Color blue each net that will make a pyramid.

a.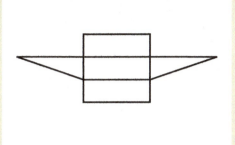

b.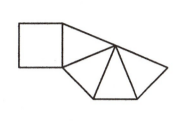

c.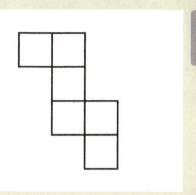

2. Solve each of these. Show your thinking.

a. $30\% \cdot b = 90$

b. $15 = 10\% \cdot r$

c. $49 = 70\% \cdot y$

d. $25\% \cdot f = 10$

Preparing for Next Year

Write a word problem to match the equation. Then find the value of the variable. Show your thinking.

$5\% \cdot p = \$200$

Working Space

Working Space

STUDENT GLOSSARY

Absolute value

The **absolute value** of a number describes the distance between the number and 0 on a number line, regardless of whether the number is positive or negative. The symbol for absolute value is | |.

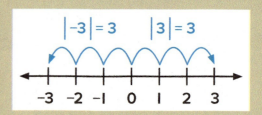

Area

The **base** (*b*) of a **triangle** can be any side. The **apex** is the vertex opposite the base. The **perpendicular height** (*h*) forms a right angle with the base and is the distance from the base to the apex. The perpendicular height can be inside the triangle (as shown on page 204) or outside (as shown on page 208). The rule (or formula) for calculating the area of a triangle is $(b \cdot h) \div 2$ or $\frac{bh}{2}$.

Parallelograms are quadrilaterals with two pairs of parallel sides. The **base** (*b*) of a parallelogram can be any side. The **perpendicular height** (*h*) forms a right angle with the base and is the distance from the base to the opposite side. The rule (or formula) for calculating the area of a parallelogram is $b \cdot h$ or bh.

The **surface area** (SA) of a 3D object can be calculated by finding the sum of the areas of each 2D shape that forms the surface. A **net** is a type of picture that shows what a 3D object would look like if it were flattened out. See page 344 for examples.

Coordinate plane

A **coordinate plane** is a rectangular grid which has a horizontal axis called the *x*-axis and a vertical axis called the *y*-axis. The **origin** is where the axes meet. An **ordered pair** is two numbers that describe a specific point on a coordinate plane. These numbers are called **coordinates**. Marking ordered pairs on a coordinate plane is called **graphing** or plotting.

The four sections of the coordinate plane are called **quadrants** and are named as shown on page 32.

Equations

An **equation** is a statement showing that two expressions are equal. To solve equations involving one variable, use the opposite (or inverse) operation. For example:

Addition	Subtraction	Multiplication	Division
$20 = g + 5$	$m - 9 = 15$	$3f = 18$	$q \div 5 = 7$
$20 - 5 = g + 5 - 5$	$m - 9 + 9 = 15 + 9$	$3f \div 3 = 18 \div 3$	$q \div 5 \cdot 5 = 7 \cdot 5$
$15 = g$	$m = 24$	$f = 6$	$q = 35$

STUDENT GLOSSARY

Exponents

Exponents are often used to represent multi-digit numbers greater than one million. It involves repeatedly multiplying a base number as many times as indicated by the exponent. For example, 10^6 is equivalent to $10 \times 10 \times 10 \times 10 \times 10 \times 10$, or 1,000,000. Any number with an exponent of 0 is equivalent to 1.

Expressions

An **expression** can be a number or letter by itself (e.g. 2 or h), or a combination of numbers, letters, and operations without a relationship between them (e.g. $2 \times h + 4$). In algebra, the multiplication symbol × can be confused with the letter x. For this reason, expressions such as $2 \times h + 4$ can be recorded as $2 \cdot h + 4$ or $2h + 4$.

In the expression $2h + 4$, the letter h is called the **variable** because its value can change. The expression also contains a **constant**, which is 4. A constant does not change value. In $2h + 4$ the number 4 is also considered a **term** of the expression. The combination of 2 and h is treated as one single term. If one factor of a term is a known number, then that factor is called a **coefficient**. The coefficient is always written before the variable. For example, $2 \cdot h$ and $h \cdot 2$ are both written as $2h$.

To **simplify** an expression, **like terms** are combined. This means all single numbers are combined, all the same variables are combined, and all the same squared variables are combined: $5 + m^2 + 4 + 2m + 3m^2 + 4m$ becomes $4m^2 + 6m + 9$.

Relationships can be described in a general way by naming the types of variables involved. **Independent variables** (or inputs) are the starting numbers. **Dependent variables** (or outputs) are the result of a change to the independent variable. Relationships can be represented in tables and on **coordinate planes**. Independent variables are usually on the **x-axis**. Dependent variables are usually on the **y-axis**.

Inequality

An **inequality** is a relationship between two quantities that are not equal e.g. $10 < 2p$. The **range of possible values** for an inequality can be shown on a number line. Consider the following statement: *The temperature on Monday was below 30°F.* The maximum possible value must be less than 30°F. This can be shown on a number line with an open circle around the 30 tick mark to indicate that 30 is not included in the range.

STUDENT GLOSSARY

Inequality (continued)

When there is a limit to the minimum possible values, the orange arrow stops at the limit and ends with a solid circle at that position. Consider the following statement: *Fewer than 20 children came to school in the storm.* The range of possible values cannot be less than 0 and it must be less than 20, so it can be shown on a number line.

The same thinking can be used to represent the range of possible values in situations where there is no limit to the maximum possible value (as shown below left), and in situations where there is a limit to the maximum possible value (as shown below right).

Order of operations

The **order of operations** indicates the sequence in which operations are to be completed when working with mixed operations. Working **left to right** along the equation:
- Perform any operation inside brackets or parentheses. When more than one pair of grouping symbols are involved, start with the innermost grouping symbols.
- Calculate any products involving exponents.
- Multiply or divide pairs of numbers.
- Add or subtract pairs of numbers.

Percentage

A **percentage** is a quantity expressed as a fraction of 100 using the symbol %. For example, if there are 100 cars and 20 of them are blue, the percentage of cars that are blue is 20% (read as *twenty percent*).

Positive and negative numbers

Any number can be **positive** or **negative**, except zero. Positive numbers are greater than zero. Negative numbers are the same distance from zero as positive numbers are, but in the opposite direction. The **negative symbol** indicates the opposite value of a number. So −4 can be read as *negative four*. It can also be thought of as *the opposite of four*. Similarly, −(−5) can be thought of as the *opposite of −5*. **Integers** are counting numbers (1, 2, 3…), zero, and the opposites of counting numbers (−1, −2, −3…).

STUDENT GLOSSARY

Ratio

A **ratio** describes a constant relationship between quantities and may involve the same units (like grams) or different units (like grams and tablespoons). Because this relationship stays constant as the quantities change, **equivalent ratios** have the same relationship between the quantities.

Ratios can be used to describe a **part-part** situation (e.g. the number of red flowers in relation to the number of white flowers in one bunch) or a **part-whole** situation (e.g. the number of red flowers in relation to the total number of flowers in the bunch). If a ratio describes a part-whole relationship, it can be written as a common fraction, decimal fraction, or percent.

A **rate** describes the relationship between two quantities when the value of one of the quantities is 1. Rates may be expressed in different ways but all have 1 as a quantity: $8 for 1 ticket, 2 cups of flour for every 1 cup of milk, 3 miles per hour, or $4/lb.

Statistics

A **measure of center** is a value that represents a typical value in a data set. The **mode** is the value that occurred most often. The **median** is the middle value when the data is arranged in order, or the mean of the two middle numbers when the size of the set is even. The **mean**, or **average**, is calculated by adding all the values in the data set together, then dividing the sum by the number of values.

The **mean absolute deviation** (MAD) is one way to look at the variation between the mean and each data point. The distance from the mean to each data point can be called the **deviation**. It is an **absolute value**. See page 358 for details on how to calculate the MAD.

The **interquartile range** (IQR) is another way to evaluate how data varies from the average. After the data set is ordered from least to greatest, the median is found, then the median of each half is found. These medians are called **quartiles**. The IQR is calculated by finding the difference between the first and third quartile.

interquartile range

| 8, 9, 10, 11, 13, 14, | 14, 14, 16, 17, 17, 18, | 18, 18, 18, 19, 19, 19, | 20, 20, 20, 22, 22, 24 |

first quartile   second quartile   third quartile

IQR = 19.5 − 14 = 5.5

A **box plot** is a graph that shows a summary of a data set. See page 364 for more details.

TEACHER INDEX

Absolute value 26–7, 36–7, 137, 331

Academic vocabulary 16, 28, 40, 54, 66, 78, 92, 104, 116, 130, 142, 154, 168, 180, 192, 206, 218, 230, 254, 266, 278, 292, 304, 316, 330, 342, 354, 368, 380, 392, 406, 418, 430, 444, 456, 468

Addition
Common fractions 143, 149, 424
Decimal fractions 29, 60, 68–9, 98, 110, 136, 142, 186, 298
Standard algorithm 29, 68–9
Three-digit numbers 10
Three whole numbers 34
Two-digit numbers 10

Algebraic thinking
Equations
  Backtracking 152–3
  Solving 135, 225, 231, 256–9, 262–6, 268–71, 273–7, 279, 400, 425, 452–3, 469
  Writing
    With one variable 122–3, 131, 134–5, 225, 231, 267
    With two variables 126–7, 152–4, 213
Expressions 120–1, 125
  Comparing 66, 320–3, 325–9, 387
  Evaluating 128–9, 137, 311, 317, 419
  Simplifying 244–7, 249–53, 255, 261, 267, 276–7
  Writing 120–1, 125, 219
Inequalities 320–3, 325–9
  Comparing 320–1, 325, 337
  Representing 322–3, 326–9, 331
Language 44, 120, 125, 130, 142, 246
Order of operations 11, 40, 46–7, 50–5, 87, 93, 99, 132–3, 180–1, 400
Patterns
  Addition 139, 141, 145, 147, 155
  Division 431
  Multiplication 11, 49, 55, 105, 111, 138–41, 144–7, 149
  Rule 138–41, 144–5
  Subtraction 49, 147
  Tables 140–1, 144–7, 149–53, 431

Algebraic thinking (continued)
Problem solving
  Think Tank problems 16, 28, 40, 54, 66, 78, 92, 104, 116, 130, 142, 154, 168, 180, 192, 206, 218, 230, 254, 266, 278, 292, 304, 316, 330, 342, 354, 368, 380, 392, 406, 418, 430, 444, 456, 468
Word problems
  Addition 270–1, 276–7
  Area 73, 77, 304
  Capacity 105, 155, 160, 191
  Common fractions 131, 137, 155, 158–61, 163–4, 168–9, 177, 287, 309, 315
  Decimal fractions 115
  Division 107, 115, 131, 158–61, 163–4, 168, 177, 274–7, 309, 315
  Length 105, 154–5
  Multiplication 59, 63, 77, 145, 187, 193, 270–1, 276–7
  Order of operations 11, 45, 47, 49, 53, 133
  Percentage 297, 301, 303, 307, 375, 409, 413, 437, 443–4, 448–9, 454–5, 463
  Perimeter 73
  Positive and negative numbers 27
  Rate 217, 220–3, 227, 231, 249, 255, 369, 418
  Ratio 89, 96–7, 100–1, 104, 179, 184–185, 188–91, 217, 249, 282, 297, 398–9, 404–7, 419, 463
  Subtraction 274–7
  Surface area 352–4
  Variables 121–3, 133–5, 250–3, 259, 263, 270–1, 274–7, 305, 454–5
  Volume 79, 279, 355, 390–2
Variables 120–3, 126–9, 131–5, 143, 150–1, 225, 231, 452–5
  Generating and graphing 458–61, 464–7

Comparing
Common fractions 9
Decimal fractions 9, 35, 287
Expressions 66, 320–3, 325–9, 387
Positive and negative numbers 24–5, 27, 35, 131, 137
Rates 228–30, 369, 380
Ratios 178–9, 182–3, 369, 402–3
Polygons 230, 410–1, 414–5, 431, 439, 445

TEACHER INDEX

Coordinate plane
  Calculating distance 36–7, 230
  Graphing 30–1, 33, 37–9, 61, 117, 146–7, 230, 431, 458–61, 464–7
  Language 30, 32, 40, 146
  Ordered pairs 30, 31, 33, 36–9, 41, 117, 146–7
  Ratios 90–1, 182–3, 190
  Reflections 38–9

Division
  Extending
    Five-digit numbers 103
    Four-digit numbers 73, 102–3, 106–7, 111, 187
    Fractions
      Common fractions 125, 131, 137, 158–61, 163–73, 175–7, 180–1, 187, 193, 279, 287, 293, 299, 308–9, 312–5, 343, 349, 355, 450
      Decimal fractions 35, 41, 102–3, 108–9, 112–5, 162, 193, 212, 219, 225, 231, 336, 412
    Three-digit numbers 22, 87, 181, 187, 286
    Two-digit numbers 22, 93, 109, 117
  Mental strategy
    Invert and multiply (common fractions) 308–9, 312–3, 349, 355
  Related to common fractions 261, 325
  Remainders 73, 106–7, 111, 117
  Repeating 108–9
  Standard algorithm 73, 93, 102–3, 106–9, 111–7

Exponents 6–7, 11, 14, 17, 23, 50–1

Fractions
  Common fractions
    Addition 143, 149
    Area model 67, 158, 160, 163–4, 170
    Comparing 9, 143
    Division 125, 131, 137, 158–61, 163–73, 175–7, 180–1, 187, 193, 279, 293, 299, 308–9, 312–5, 343, 349, 355, 450
    Equivalent 55, 143, 305
    Improper fractions 149
    Mixed numbers 149
    Multiplication 17, 255, 260, 267, 273, 362

  Negative 18
  Related to decimal fractions 284
  Related to division 261, 325
  Related to ratio 282–5, 293
  Subtraction 424
  Decimal fractions
    Addition 29, 60, 68–9, 98, 110, 136, 142, 186, 298
    Area model 23
    Comparing 9, 35, 130, 287
    Division 35, 41, 102–3, 108–9, 112–5, 162, 193, 212, 219, 225, 231, 336, 412
    Hundredths 124
    Multiplication 11, 17, 23, 29, 48, 61, 67, 70–1, 74–7, 79, 201, 207, 213, 224, 248, 438
    Negative 18–21
    Rounding 12–3
    Subtraction 68–9, 72, 124, 136, 142, 272, 348

Integers 18–21, 24–9, 337

Measurement
  Area
    Composite shapes 169, 175, 278, 292
    Parallelogram 196–9, 206–7, 325
    Polygon 214–5, 218, 225, 317, 337, 416–21
    Quadrilateral 210–1, 219, 354, 381
    Regular shapes 104, 163, 192, 196–7, 230
    Surface area
      Prism 344–5, 350–4, 426–9, 451, 463, 468–9
      Pyramid 346–7, 350–4, 422–3, 427–9, 469
    Triangle 202–5, 208–9, 213, 305, 311, 331, 381
  Capacity 99
  Mass 111, 117, 254
  Perimeter
    Irregular polygons 278, 292
    Regular polygons 104
  Volume
    Composite prisms 273, 378–9, 387
    Rectangular-based prisms 116, 130, 267, 304, 378–9, 382–5, 388–93, 401, 407
    Rule for calculating 378–9
    Unit cubes 349

TEACHER INDEX

Multiplication
- Comparison model (tape diagram) 49
- Convention 44
- Distributive property 64–5, 73, 78, 201
- Estimating 70–1, 74–5, 77
- Exponents 50–1, 175
- Extending
  - Four-digit numbers 41
  - Fractions
    - Common fractions 17, 255, 260, 267, 273, 362
    - Decimal fractions 11, 17, 23, 29, 48, 61, 67, 70–1, 74–7, 79, 201, 207, 213, 224, 248, 438
  - Three-digit numbers 35, 41, 324
  - Two-digit numbers 41, 86, 148, 174, 200, 374
- Factors 23, 56–7, 62–6, 193, 207, 462
- Mental strategies
  - Double and halve 61, 86, 374
  - Partial products 148
  - Round and adjust 67
  - Use a known fact 148
- Multiples 56–9, 67, 187
- Patterns 11
- Prime and composite numbers 56–7
- Standard algorithm 35, 41, 70–1, 74–7, 79

Number line 18–21, 24–7, 29, 58, 284, 290–1, 293–6, 306, 322–3, 326–9, 330–1, 358–9, 363–7, 375, 393, 413, 439

Number representation
- Common fractions 8
- Decimal fractions 8
- Multi-digit numbers (>1,000)
  - Abbreviations 12, 17, 163, 169
  - Expanded form 6–7, 11
  - Exponents 6–7, 11, 14
  - Symbolic 12, 17
- Positive and negative numbers 18–21, 125, 137, 293, 331, 439

Ordering
- Positive and negative numbers 24–5

Ratio
- Comparing 178–9, 182–3, 369, 402–3, 468
- Concept 82–3, 87, 105
- Equivalent 84–5, 88–9, 93–5, 99, 155, 287, 363
- Graphing 460–1, 464–5
- Interpreting 96–7, 155, 192, 249, 266
- Language 82–3
- Models
  - Area 284–5
  - Coordinate plane 90–1, 117, 182–3, 190
  - Table 88–90, 117, 178, 182–3, 190–2, 283, 287, 396–7
  - Tape diagram 82–5, 93, 96, 184, 188, 398, 402–3, 413
- Percentage
  - Calculating 294–7, 300–7, 310–1, 317, 363, 368–9, 375, 386, 407–9, 413, 418, 436–44, 446–9, 451–5, 457, 469
  - Complementary 440–1
  - Greater than 100% 434–7, 439, 442–4
  - Language 288
  - Models
    - Area 288–9, 299, 401, 434
    - Number line 290–1, 306
  - Related to ratio 291, 305, 437, 439
  - Resizing 2D shapes 408–9, 416–7, 420–1, 425, 451, 457
- Rate
  - Calculating 220–3, 226–7, 231, 249, 418, 445
  - Comparing 228–9, 261, 369, 380
  - Concept 216–7, 220–1
- Related to fractions 282–5, 293
- Sharing in a given ratio 184–5, 188–9, 398–9, 406
- With three parts 396–9, 401–6, 419

Rounding
- Decimal fractions 12–3
- Numbers > 1 million 169

Shape
- Three-dimensional objects (attributes) 387, 393
- Two-dimensional shapes (comparing) 410–1, 414–5, 418, 431, 439, 445
- Angle 206, 218